ABHANDLUNGEN
DER AKADEMIE DER WISSENSCHAFTEN DER DDR
Abteilung Mathematik, Naturwissenschaften, Technik

Jahrgang 1977 · Nr. 8 N

SIEGFRIED DÄHNE AND SIEGFRIED KULPE

Structural Principles of Unsaturated Organic Compounds

With special reference to X-ray structure analyses of coloured substances

Mit 17 Tabellen

AKADEMIE-VERLAG · BERLIN · 1977

Herausgegeben im Auftrage des Präsidenten
der Akademie der Wissenschaften der DDR
von Vizepräsidenten Prof. Dr. Heinrich Scheel

Erschienen im Akademie-Verlag, 108 Berlin, Leipziger Straße 3—4

Lizenznummer: 202 · 100/496/77
Gesamtherstellung: VEB Druckhaus „Maxim Gorki", 74 Altenburg
Bestellnummer: 762 461 1 (2001/77/8N) · LSV 1275
Printed in GDR
DDR 24,— M

On the occasion of the centennial of the formulation of the first dye theory by O. N. Witt and the 50th anniversary of the publication of the conception of the chromo-state and the polymethines worked out by W. A. Ismailsky and W. König we dedicate this work to the countless scientists everywhere in the world who in the past one hundred years have contributed by their unceasing efforts and many ingenious ideas to the development of the dye theory.

Siegfried Dähne and Siegfried Kulpe

Berlin, Spring 1976.

Preface

by *Professor Dr Wolfgang Schirmer*

Many of the advances in our knowledge of the structure of chemical compounds have been brought about by the interaction of experimental findings and their theoretical interpretation. This is true of problems relating to optically active compounds as well as of tautomerism or resonance phenomena. The model of a polymethine structure presented by the authors in this work is characterised by a chain of atoms occupied with a π-electron system whose number of electrons differs from the number of links by plus or minus one. Hence polymethine compounds represent a bond type distinct from the aromatics and the polyenes.

The authors verify the usefulness of this concept by applying it to several hundred organic compounds. In the present case, too, the model was developed only after extensive experimental data had been available which enabled bond lengths and bond angles to be determined with a high degree of accuracy. It is no doubt owing to the consistent use of X-ray diffraction methods that these experimental data are available now, although we must not underestimate the importance of spectroscopic techniques in structure determination.

Considering the development of the concept of structure in chemistry, the model of the polymethine state is presented comparatively late. A hundred years have passed since O. N. Witt formulated the first theoretical approach for interpreting the structure-property relationship of organic dyes. In 1976 we will be able to celebrate the 50th anniversary of the first fundamental work on the polymethines written by W. König.

The polymethine concept by no means applies only to dyes. It seems to carry all the prerequisites enabling it to become a model representation on a par with the structures of the aromatics and polyenes. It is proving its efficiency today in that it is leading to the posing of problems for future studies.

If the structural problems mentioned in the concluding chapter of this work can be solved successfully using the polymethine model, nothing is likely to stand in the way of a universal recognition of this obviously sound concept. Even now it should however be used for studying structural elements of all relevant compounds from this new point of view and for finding more economical ways of determining these.

Professor Dr Wolfgang Schirmer

Acknowledgements

The assistance of Academician *Professor Dr Wolfgang Schirmer*, Director of the Central Institute for Physical Chemistry of the GDR Academy of Sciences, whose interest greatly encouraged the progress of this work, and of *Professor Dr Klaus Junge*, Director of the Central Institute for Optics and Spectroscopy of the GDR Academy of Sciences, is gratefully acknowledged.

Many colleagues have aided greatly with their helpful discussions and many interesting suggestion Particular thanks are due to *Dr Jürgen Fabian* and *Dr Horst Hartmann*, Technische Universität Dresden, GDR, and *Dr Douglas Smith*, Research Laboratories, Eastman-Kodak Co., Rochester, U.S.A.

We owe a debt of especial gratitude to our co-workers, Frau *Barbara Bornowski*, Frau *Freia Moldenhauer* and Herr *Burkhard Schulz*, for their untiring efforts in sifting many thousands of original contributions. Last, but not least, we are indebted to Frau *Hannelore Herda*, Leipzig, for the great trouble she has taken over the translation.

Contents

The representatives of organic chemistry are coming more and more to the conclusion that the formulae of limiting states which we have been using so far fail to reflect the real conditions prevailing in nature, for there exist such subtle differences in the state of matter compared to which our methods of description that still are very simple sometimes appear wholly inefficient.

The restless minds of most researchers feel satisfied as soon as theories have been brought forth which by and large correspond with the facts. On the basis of these theories, then usually follows a quieter period of development, a period, in which generally more facts are being collected until once again one is agreed that these theories are no longer tenable.

That has always been so, even to this day.

W. A. Ismailsky
Thesis, Technische Hochschule Dresden, 1913.

1. Introduction

Since 1965, the 100th anniversary of the formulation of the valence structure for benzene by A. Kekulé, the concept of the aromaticity of organic compounds has been repeatedly called in question [1, 2, 3], one of the reasons being coloured organic compounds such as the metal chelates of acetylacetone [4, 5] or the diazepinium salts [2], which by their chemical and physical properties such as electrophilic and nucleophilic reactivity, for instance, apparently behave like aromatics without having the ring structure with a completely delocalised $(4n + 2)$ π-electron system typical for aromatics.

These difficulties may be avoided if one extends the well-known concept of the existence of an energetically stabilised aromatic state and a polyene-like state by that of an energetically stabilised polymethine state [6, 7, 8]. Thus the deep colour of many organic compounds need not be attributed to the effects of substituents in aromatic and polyenic compounds as it has been common practice thus far.

The deep colour[1] is inherent in the ideal polymethine state. By this approach, any given intermediates, actually existing, between the three ideal states may account for the diversity and variability of the electronic structure of organic π-electron compounds. The principle determining a given structure is the relationship between the resonance (delocalisation) energies of cyclic, aromatic structural groups and chain-shaped polymethinic structural groups. This approach provides a heuristic system of unsaturated organic compounds based on their physico-chemical properties. We proposed to call this system triad system [9].

The characteristic electronic structure of a largely ideal, aromatic compound is realised in benzene (1.1); and in the polyenes (1.2) that of a nearly ideal polyenic compound.

(1.1) [10]

$$R-(CH{=}CH-)_n R \quad (1.2)$$

[1] "Deep colour" will be used in the sense of an intense light absorption at relatively long wavelengths when comparing substances with an equal number of π-electrons.

Characteristic features of **aromatic** compounds are

1. largely equalised bond lengths, resulting in the case of benzene in identical bond lengths of 1.39 Å,
2. identical π-electron densities on all atoms, which normally equal unity and which account for the sp^2 hybridisation of carbon atoms with a bond angle of 120°,
3. high delocalisation energies and correspondingly high resonance energies, ranging up to 36 kcal per mole and per π-electron sextet, respectively, in the case of benzene
4. very low chemical reactivity.

Typical features of ideal **polyenes** are

1. alternation of bond lengths between the carbon atoms of the polyene chain, ranging from 1.32 up to 1.48 Å [11–12]. The example in (1.3) shows the structure of the retinal in all-trans con figuration which is responsible for visual perception.

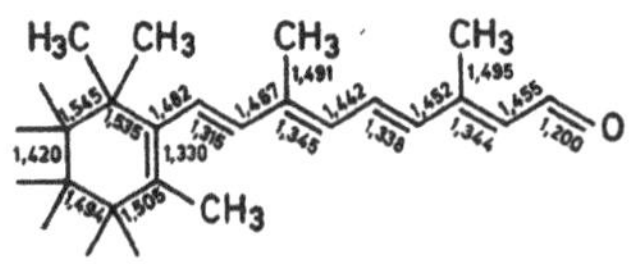

(1.3) [15]

2. similar to the aromatics, identical π-electron densities which are equal to unity normally, on all atomic centres.
3. strong tendency to chemical addition reactions.

According to HÜCKEL's theory, carbocyclic compounds with $(4n + 2)$ π-electrons distributed over $(4n + 2)$ carbon atoms should come closest to the ideal aromatic state, whilst the so-called 'anti-aromatics' with $4n\,\pi$-electrons on $4n$ atomic centres should have a more polyene-like structure. These predictions were largely confirmed by X-ray structure determinations of aromatic [10]-, [14]- and [18]-annulenes [23–32] and the anti-aromatic [8]- and [16]-annulenes [33, 34]. For instance, one can compare the typical polyenic structure of tetramethylcyclooctatetraene (1.4) with the structure of bezene (1.1) and the nearly aromatic 1,6-methano-[10]-annulene-2-carboxylic acid (1.5). For the structures of other carbocyclic substances of polyenic structure refer to [18, 35–38].

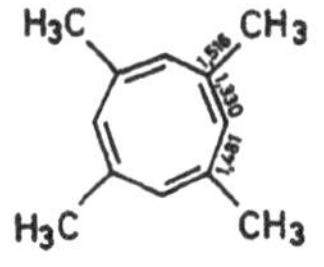

(1.4) [34]

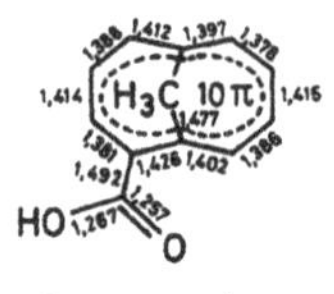

(1.5) [31]

The **intermediates** between the ideal aromatic state and the polyene state have been mainly studied experimentally by CLAR et al. [39, 40] and TROTTER [41], and theoretically by POLANSKI et al. [42–46]. The results of this study may be interpreted as follows:

In unsaturated carbo- and heterocyclic systems a maximum number of independent aromatic π-electron sextets is realised. Branching of the conjugation is avoided as far as possible in such a way that the remaining bonds assume a polyenic character. In those cases where branching is inevitable for reasons of symmetry, the ideal aromatic character is weakened accordingly.

Examples include triphenylene (1.6), phenanthrene (1.7), pyrene (1.8) and perylene (1.9), the structures of which can be deduced according to these simple rules and are well established experimentally. The bond lengths in those rings which incorporate most of the features of an ideal aromatic state average 1.39 Å; this bond length is obviously disturbed in the case of perylene (and similarly, in naphthalene and comparable carbocycles) owing to symmetry-conditioned branching of the π-electron system. Considerably greater deviations towards polyene-like single or double bonds occur in all other rings,

(1.6) [47]

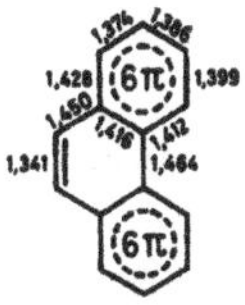

(1.7) [48]

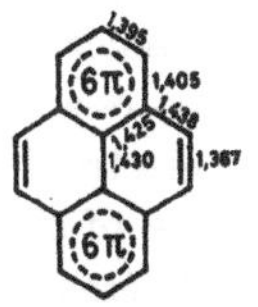

(1.8) [49]

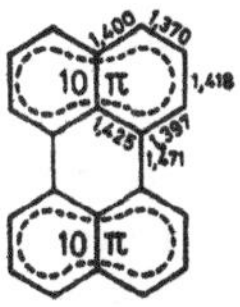

(1.9) [50]

which do not permit of formulating an isolated π-electron sextet according to the above rules.

The heptafulvalenes (1.10) and (1.11), of which very precise X-ray analyses are available, are another example demonstrating that the tendency of aromatisation and a branching of the conjugation strongly affect the bond lengths. The simple fulvalene (1.10) is polyenic in structure. Its bond lengths deviate slightly from the theoretical

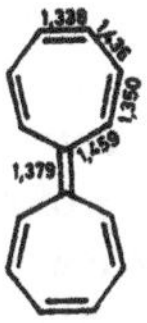

(1.10) [51]

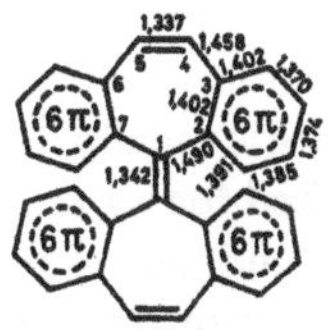

(1.11) [52]

1.33 Å and 1.47 Å, pointing to a conjugation within the ring system. At 1.38 Å the double-bond distance between the rings is relatively long. In the tetrabenzofulvalene (1.11) the four aromatic π-electron sextets dominate, thereby enlarging the double-bond distances in the 2,3- and 6,7-positions to 1.40 Å. But the 1,2-(7,1-) and 3,4-(5,6-) positioned formal single bonds are likewise caused to be enlarged by the favoured resonance within the benzene rings, the result being that the central double bond is obviously less involved in the overall resonance of the molecule, thus reducing its bond length to nearly that of a double bond of 1.33 Å.

In polyene-like carbocycles, the alternation in bond lengths, which is typical of the polyene state, may be partly eliminated by π-complex bonding with metal atoms from subgroup elements or by the formation of ions of the carbocycles. This leads, in part, to quasi-aromatic bond lengths [53—59] as can be seen in the example of bis(cyclo-octatetraene)-iron (1.12) or the dipotassium-cyclooctatetraenide [53] as compared to the complex-free

(1.12) [56]

compound (1.4). This way it is possible to stabilise unstable carbocycles like the anti-aromatic heptalene [60]. A more exhaustive study was made into the structures of complexed azulenes, which will be dealt with in detail under Chapter 8. The bonds of linear polyenes may be affected in a similar way by complexing [61—63]. No systematic study has been made thus far into the possible effects of complex formation on the light absorption of the molecules.

1.1. *X*-ray Structure Determination and Structure Chemistry

Whether or not structure-chemical conclusions are tenable is a question of the degree to which experimental results can be interpreted by or made to correspond approximately, at least, with theoretical models or concepts. The conception of the existence of aromatic, polyenic and polymethinic states of unsaturated organic compounds relates especially to assumptions on the electronic structure of molecules. A given electronic structure of a molecule, however, corresponds to a defined spatial configuration of the atoms in the molecule, quite specific interatomic distances and valence angles. There is a close correlation between the electronic structure and the geometry of the molecule.

Experimental findings on the electronic structure of molecules are gained by various spectroscopic techniques (*IR*, *UV*, *NMR*, *ESR* ...). Experimental knowledge on the geometry of molecules is obtained by means of diffraction techniques, in particular with the help of *X*-ray analysis. In simple and favourable cases spectroscopic methods too will permit conclusions to be reached with regard to the geometry of a molecule. If, however, we are concerned with experimental data on the complete geometry of larger molecules, — and only such will be of interest here —, then diffraction methods will surpass spectroscopic techniques; the main reason being that in spectroscopic methods the number of measurable experimental data is usually far less than the number of molecular parameters to be determined and required for an unambiguous determination of the geometry of a molecule. The number of experimental data in diffraction methods is generally many times greater than the number of molecular parameters to be determined. This enables the geometry of the molecule to be ascertained with an accuracy adequate for structure-chemical interpretations.

At present, *X*-ray analysis is the only method which enables all details of the complete spatial structure of the molecules in the crystal lattice to be determined. Where structure-chemical considerations are based on structure-analytical findings, the accuracy of *X*-ray structure determinations is of eminent importance. The difference in the interatomic distance between a $C_{sp}—C_{sp}$ single bond of 1.37 Å and that of a $C_{sp^2} = C_{sp^2}$ double bond of 1.35 Å, for instance, is of relevance in terms of structural chemistry. To realise by structure analysis this difference of 0.02 Å means that the triple standard deviation 3σ must be reduced to less than this value. Consequently the single standard deviation must be $\sigma < 0.007$ Å. To secure such a limit of error for all interatomic dis-, tances of a molecule in a crystal cell requires a maximum of effort, equipment and care, especially in the case of loose-packed organic molecules.

Not every *X*-ray analysis published was carried out with the intention of determining precise interatomic distances. Often it serves to clarify questions relating to conformation or configuration which to resolve does not necessarily require the most accurate interatomic spacings. If we want to estimate the relevance to structure chemistry of the X-ray analytical results published, we need to pay great attention to the aspect of accuracy. It is particularly useful to determine the averages of geometric parameters (interatomic spacing ,valence angle, etc.) of different, independently determined structures which are structure-chemical equivalent. The standard deviations obtained from accurate X-ray analyses suffice for essential conclusions on bonding and electronic structures of molecules to be derived from their geometry. The accuracy is also adequate for examining the soundness of quantum-chemical calculations and, what is more important, for serving as initial data for such calculations.

X-ray analysis not only provides accurate data on the required bond lengths but also on the bond angles which are determined by their spatial arrangement and by the nature of the hybridisation of the atoms. It is well-known that the amounts of energy needed to effect changes in bond angles are smaller by one order of magnitude than those required to produce comparable alterations in the interatomic spacings. For example, changes in hybridisation as a function of charge density could be proved in methine atoms [64] which had not been predicted theoretically. A detailed treatment follows under Section 3.1.

Every step towards more accurate X-ray analyses is of great importance, particularly for finding out experimentally established correlations between structure and bonding in the molecular and intermolecular fields. There has been evidence that there is further scope for the use of X-ray analysis in combination with neutron diffraction. There is, moreover, a growing number of structure determinations, in which the positions of the hydrogen atoms in the structure can be determined conclusively from X-ray diffraction data. The fact that generally the positions of the hydrogen atoms are less accurate by one order of magnitude than those of the non-hydrogen atoms, is of secondary importance for often much is gained already by knowing the absolute or relative positions of the H-Atoms in a structure.

In the present work the emphasis is mainly on interatomic distances and how these relate to structure-chemical features. As a rule it will not be possible to discuss in detail valence angles, planarity of parts of a molecule or whole molecules, the angles between such planes (dihedral angle), the distances between cations and anions, the distances between adjacent molecules and other factors which often are essential in assessing details.

X-ray analysis carries one limitation however. It can only investigate molecular structures of compounds in the crystalline state. Usually even single crystals are necessary for X-ray determination. This restriction brings up the question — to what extent may the molecular structure in crystals be considered valid for the liquid or the gaseous state. Extensive experimental findings on simple inorganic and organic substances have shown, however, the geometry of many molecules in the crystal lattice to be nearly identical with that in the liquid and in the gaseous state. Modifications, if any, can be frequently determined by comparing X-ray data with the results of other analytical methods, e.g. spectroscopic techniques. The following review describes only such molecular geometries as determined by X-ray analysis despite the fact that in many cases, especially with the simple polymethine dyes, a large number of spectroscopic results are available. We generally proceed from the assumption that the molecular structure in the crystal lattice closely resembles the structure of the free molecule, unless explicitly stated otherwise.

The existing literature on X-ray analyses of unsaturated organic molecules is extensive and is being constantly enlarged, providing a growing body of valuable information as the accuracy of experiments and methods is increasing [65—70]. To date, there has been no comprehensive presentation on the X-ray structures of coloured organic compounds.

The present work is an attempt to evaluate and systematise these structures on the basis of the polymethine concept. The discussion covers mainly the X-ray analyses obtained between 1960 and 1975. Preference is given especially to publications of Acta Crystallographia in the past three years. The compilation does not claim to be complete. Works presenting inaccurate results — or such, whose accuracy cannot be jugded reliably, have been omitted where possible. In some cases, however, we cited also structures which are not very accurate purely for the sake of expounding the new approach. We are aware of the fact that we have also omitted many interesting structural data which may serve as a further proof of the developed concept of the three ideal states. In some cases structural data of characteristic key substances are missing which may stimulate fresh structure investigations. But there may be also entities which seem to be exceptions to the triad system and which should challenge further investigations into the relations between colour and constitution of unsaturated organic compounds.

2. Structural Features of Coloured Organic Compounds

After fundamental preliminary work by ISMAILSKY [71, 72] and W. KÖNIG [73] the concepts of the ideal polymethine state (chromo-state) were first empirically derived by DÄHNE [74] using extensive experimental results and later, in collaboration with LEUPOLD and NOLTE [6, 7], from quantum-chemical calculations based on a completed HÜCKEL Molecular Orbital method. Recently FABIAN and HARTMANN [76, 77] showed that the model of the polymethine state follows conclusively from the simple *HMO* model. By suitable sectioning of unsaturated cyclic hydrocarbons into chain fragments it is possible to represent the polymethinic and polyenic structures, their energies and molecular parameters, such as π-bond orders and π-electron densities which can be deduced from the wave-functions, in the form of analytical functions, in the same way as it is possible with aromatic hydrocarbons [78, 79].

It was found [74] that most coloured organic compounds contain polymethinic fragments of the general formula (2.1). It is a typical feature of these structural groups that a chain of N atoms X, X' and Y, mainly from groups IV, V, or VI of the periodic table, is occupied by a system of π-electrons whose number is smaller or greater by **one** than the number of atoms in the chain.

$$\begin{array}{ll} \cdots\cdots (N \pm 1) \cdots\cdots & N = 2, 3, 4 \ldots \\ X - (Y)_{N-2} - X' & X, X', Y = \text{atoms of groups IV, V and VI of the periodic table} \end{array} \qquad (2.1)$$

There may be an odd or even number of atoms in the chain. In the former case, we are concerned with polymethine dyes, the best-known representatives of which are the cyanines and oxonols. In the simplest cyanine dyes (2.1) with $Y = \mathrm{CH}$; $X = X' = NR_2^{\frac{1}{2}\oplus}$; $N = 3, 5, 7, \ldots$, the ideal polymethine state is nearly reproduced.[1] Dyes with an even number of chain atoms contain necessarily an odd number of π-electrons. Consequently these are organic radicals which were termed polymethine radicals [74].

Nearly ideal polymethine dyes have a higly symmetrical distribution of the π-electrons through the chain, a maximum π-electron delocalisation and hence, a high resonance energy [80—82], high oscillator strengths (f-values) in conjunction with relatively long

[1] In textbooks and in the original literature such and similar dyes used to be classified as donor (D) and acceptor (A) substituted polyenes $D(-\mathrm{CH{=}CH})_n-A$ and substituted aromatics $D-$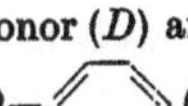$(-\mathrm{CH{=}CH})_n-A$ respectively, with $D = -\overline{\mathrm{N}}R_2$, $A = -\mathrm{CH{=}}\overset{\oplus}{\mathrm{N}}R_2$ and $n = 0, 1, 2, \ldots$

wavelength absorption bands [83—85], a very high optical polarisability [86, 87], a maximum equalisation of bond orders (corresponding to an equalisation of bond lengths, as will be shown in Section (3.1) and pronouncedly alternating π-electron densities in the ground state. The appropriate analytical model functions for polymethines with $(N + 1)$ π-electrons on N centres are [76]

$$p_{r,s} = \frac{1}{N} \operatorname{cosec} \frac{\pi}{2N}$$

for the bond orders in the ground state $p_{r,s}$ between centres r and s of neighbouring atoms and

$$q_r = 1 - \frac{(-1)^r}{N} \operatorname{cosec} \frac{\pi}{2N} (2r - 1)$$

for the π-electron densities in the ground state q_r at the centres $r = 1, 2, \ldots N$.

The electron density alternation through the polymethine chain results in an extremely high electrophilic and nucleophilic reactivity of the polymethine dye stuffs [9, 88, 89]. Also typical of ideal polymethines is the minimum change in geometry relative to bond lengths and a maximum electron density redistribution in the molecule into the first excited singlet state [7, 90, 91] when irradiated by light while the mirror symmetry of the electron density distribution along the polymethine chain is maintained.[1]

Even if the terminal atoms in (2.1) are replaced by other hetero-atoms of groups IV, V and VI of the periodic table a still approximately polymethine-like structure with the above properties will be obtained provided the π-electrons are distributed symmetrically throughout the polymethine chain. A highly symmetrical π-electron distribution is found in all dyes with identical terminal groups ((2.1) $X = X'$) when the molecules are in the singly charged, ionic state. There are, however, examples of electro-neutral dyes in the ideal polymethine state, where a symmetrical π-electron distribution is enforced by non-uniform X, X' terminal groups of a comparatively strong electron donor and acceptor effect [74]. Hence, contrary to the widely held opinion, the ionic charge is not a necessary condition for the formation of deeply coloured, unsaturated organic compounds.

The polymethine radicals containing an even number of methine groups (2.1, $N = 2$, $4, 6 \ldots$), and thus an odd number of π-eletrons, from a theoretical point of view [6], should also be deeply coloured, and on the strength of the *HMO* model for the ideal polymethine state, should likewise feature an ideal equalisation of bond lengths [76, 77]. The latter remains to be proved in the future because of the lack of suitable chain-shaped radicals for comparison.

Of primary importance for structural information are the geometric properties of the polymethines. According to quantum-chemical calculations [6, 75—77], the carbon-carbon distances in polymethine-like structural groups should be identical among one another and equal to the distances of cyclic aromatics, ideally similar to those of benzene (1.39 Å) which, in fact, has been established experimentally (Chapter 3.).

[1] We have here a maximum intramolecular charge transfer comparable to standing waves but no transport of electron charges from one end of the molecule to the other, in the sense of the traditional charge-transfer idea [92, 93].

In contrast with ideal aromatics, polymethines show a strongly alternating π-electron density distribution which has been proved only indirectly thus far. A first confirmation of the effect was provided by the chemical shifts in the ^1H- [94, 95] and ^{13}C- [96, 97] nuclear magnetic resonance spectra whose values showed a pronounced alternation. Moreover, the charges on the methine atoms cause a high polarity of the C—H bonds which affects both the characteristic bending frequencies [98] and the intensities of the C—H valence vibrations [99] in the infrared spectra of polymethine-like compounds. An at first unexpected effect was the valence angle alternation at the methine atoms of polymethines. This effect was derived by KULPE et al. [64] from precise *X*-ray studies of crystals. It is causally connected with the rehybridisation of the σ-electrons within the polymethine framework caused by π-electron density alternation [64, 100] and described in detail under Section 3.1. Apart from this, peculiarities in the intermolecular interaction effects, which favour the formation of dye aggregates [101—104], may be caused both by the charge density alternation and the very high optical polarisability of the polymethines [86, 87]. Finally, there occur anomalies with respect to the bond lengths in coupled polymethines (Sections 6.1; 6.2), which may be attributed to electron density alternation.

Like intermediates between aromatics and polyenes the intermediates between a polymethinic structure (2.2) and polyenic structures as in the formulae (2.3) and (2.4) should be identifiable by *X*-ray analyses, especially because of the growing bond length alternation.

$$\cdots\cdots (2n+4)\pi \cdots\cdots$$

$$\overline{X}(-CH{=}CH)_n-CH{=}X' \leftrightarrow X(-CH-CH)_n-CH-X' \leftrightarrow X({=}CH{=}CH)_n{=}CH-\overline{X}'$$

(2.3) (2.2) (2.4)

Electronegativity:

$X < X'$ $X = X'$ $X > X'$

The ideal, symmetrical polymethine structure with complete π-electron delocalisation is represented in (2.2) and further on by dotted lines. With reference to bond length equalisation, this structure could also be represented by using the two energetically equivalent, mesomeric limiting structures (2.3) and (2.4). This has been general practice in the textbook literature so far, but neglects the pronounced π-electron density alternation which is an immanent feature of the ideal polymethine state. As a characteristic property of the polymethine structure, the alternation needs to be taken into consideration in every case. It can be introduced into formula (2.2) by the symbols $\delta^\oplus$ and $\delta^\ominus$ as demonstrated in (2.5) in the concrete case of a simple merocyanine.

According to another suggestion [8] circles of different size are used in (2.6) for representing the π-electron density alternation. This has the advantage that the alternation

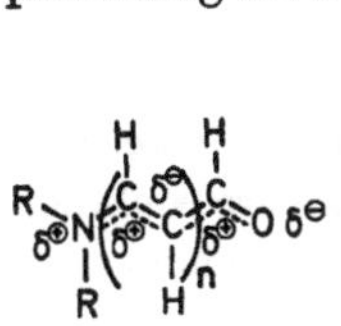

(2.5)

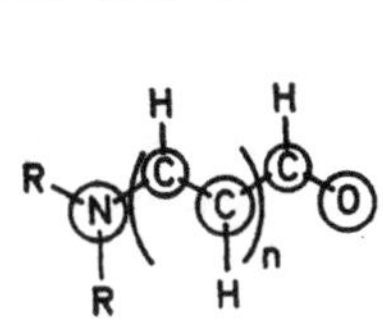

(2.6)

is readily identified even when, through substitution, a positive charge is localised on a heteroatom with a comparatively high π-electron density as is the case on the nitrogen atom of merocyanine (2.5).

In the formalism of resonance theory charge density alternation may, in principle, be represented by using additional, polar limiting structures. For a trimethine this would yield apart from (2.3) and (2.4), the additional structures (2.7) to (2.9) which, however, is not very convenient and, as is well known, allows some arbitrariness [8, 9] in rating the

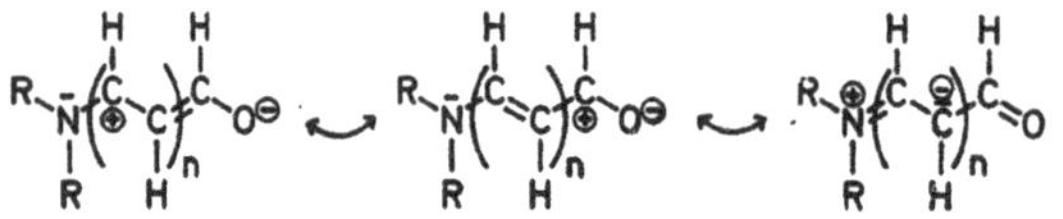

(2.7) (2.8) (2.9)

role of the weight factors in the structural resonance as a whole. As it was shown by KULPE [105], the limiting structures (2.7) to (2.9) must be taken into consideration, especially to describe the specific character of the C=O bond length found by X-ray analyses (cf. Chapter 3.) and to avoid wrong conclusions with regard to possible variation of molecular geometry in the liquid or gaseous state. In spite of this special problem, for these cases also we will use formula (2.2) in a qualified sense that it overestimates the polymethinic character. Also in the interpretation of the structure of coupled polymethines (cf. Sections 6.1; 6.2) polar limiting structures such as (2.7) up to (2.9) have considerably facilitated understanding of the experimental structural peculiarities of these compounds [106–108].

In the present work, too, two or more limiting structures are used in some instances to describe the electronic structure of a compound in the sense of resonance theory, introducing the polymethinic fragment (2.1) or (2.2), respectively, as an independent valence bond structure.

The realisation of intermediates between polymethinic structures and aromatic structures is determined by the order of magnitude of the resonance energies of the competing structural fragments. Since the average bond lengths are approximately the same in both ideal states, the changes in geometry to be expected will be slight. Here the same rule [8] applies that every branching in a linearly conjugated π-electron system will necessarily lead to a weakening and hence, to a lengthening, of the bonds concerned. As a result π-electron systems avoid branching where possible. If, therefore, preferential stabilisation of an aromatic ring system occurs in a dye molecule containing polymethinic fragments, lengthening should be more pronounced in the polymethine bonds which do not form part of the ring system than in the bonds inside the ring, and vice versa. This rule of branching, which has been previously deduced by us qualitatively in dyes [8], seems to have general application. Concerning bonding in metals [68, 109], PAULING found the sum of bond orders around each atom to be approximately constant and independent of the precise structure of the environment. BÜRGI [110] successfully used this empirical rule recently to interpret distances in inorganic complex ions and in several organic compounds not branched by conjugation. BÜRGI proved the rulès suitability as a basis for a comprehensive theoretical description of bonding and structure in inorganic and organic substances, enabling even information on reactivity and reaction mechanisms to be gained under certain conditions. Hence we are confident that, on closer examination, the rule that the sum of bond orders at an atomic centre

Table 1: Characteristic Bond Lengths [Å] In Organic Molecules*)

Bond	Length	Ref.	Bond	Length	Ref.
$C_{sp^3}-C_{sp^3}$	1.53	[111–114]	C–S	1.81	[68]
$C_{sp^3}-C_{sp^2}$	1.51	[111–114]	C=S	1.56	[70, 116]
$C_{sp^3}-C_{sp}$	1.46	[111–114]	C–F	1.37	[68]
$C_{sp^2}-C_{sp^2}$	1.47	[111–114]	$C_{sp^3}-Cl$	1.76	[111]
$C_{sp^2}-C_{sp}$	1.43	[111–114]	$C_{sp^2}-Cl$	1.72	[111]
$C_{sp}-C_{sp}$	1.37	[111–114]	$C_{sp}-Cl$	1.63	[111]
$C_{sp^2}=C_{sp^2}$	1.33	[111, 113, 114]	N–O	1.36	[69, 70]
$C_{sp^2}=C_{sp}$	1.31	[111, 113, 114]	N=O	1.23	[69, 70]
$C_{sp}=C_{sp}$	1.28	[113, 114]			
$C_{sp}\equiv C_{sp}$	1.20	[67, 111, 113, 115]	N–N	1.46	[69, 70]
$C_{sp^3}-O$	1.43	[111]	N=N	1.24	[69, 70]
$C_{sp^2}-O$	1.37	[111]	N≡N	1.10	[69, 70]
$C_{sp^2}=O$	1.21	[111]			
$C_{sp}=O$	1.17	[111]			
C–N**)	1.47	[70, 111]	S–S	2.10	[115]
C=N	1.30	[111]	S=S	1.89	[115]
C≡N	1.16	[70, 111]			

*) The standard bond lengths discussed in the following sections are not always consistent with the values in the references but generally inside these values by ±0.01 A. The data cited herein are taken from diffraction and spectroscopic experiments. The carbon-carbon bond lengths hold mainly for open-chain molecules. Greater deviations from these standard values occur in particular in molecules containing cyclic groups.

**) The C–N bonds have not been classified by hybrid state. Thus it appears to be difficult to decide on a characteristic C=N double bond, for instance [70, 69, 111].

is constant will furnish satisfactory quantitative results for branching of conjugated systems.

Finally, turning from a polymethinic to an aromatic and/or polyenic structure should be accompanied by a diminishing of all those effects caused by the characteristic charge density alternation and the high polarisability of the polymethines.

To assess experimentally determined bond lengths in terms of structure chemistry, based on the rules derived above, requires a comparison with the standard bond lengths of pure single, double or triple bonds. These are summarised in Table 1. The bond lengths between atoms in a homopolar compound depend considerably on the degree of hybridisation of the atoms constituting the compound as well as on the portion of single or multiple bond-character of the bonds concerned.

The following general rules apply:

1. as the degree of the *s*-character of hybridised bond orbitals increases, the bond lengths become shorter without, however, leading to a substantial increase in the portion of double or multiple bond-character;
2. as the portion of double or multiple bond-character increases, the bond lengths become shorter without affecting the degree of hybridisation.

Other structure-chemical factors influencing bond lengths are hyperconjugation effects, differences in electronegativity, the electric charge of the constituent atoms, intramolecular steric interactions as well as intermolecular interactions, of which the latter two have a minor influence normally on the bond lengths. To decide which effects are mainly responsible for the bond lengths often presents considerable difficulties and is possible only within the structure-chemical context.

The recommended values given in Table 1 should be taken merely as approximate values for assessing the structural properties of organic molecules. Some of the values need yet to be determined more exactly. In particular it would be desirable to ascertain more detailed values as a function of the hybrid state for some typical single or multiple bonds which are of relevance to organic chemistry such as, for example, the carbon-nitrogen bonds.

3. The Structure of Nearly Ideal Polymethine Dyes

3.1. Cyanines

Systematic structure analyses of the simplest vinylene homologue polymethine-cyanines (3.1) have been carried out by KULPE and co-workers [117—124]. The results are summarised in Table 2. The structures of some meso-substituted monomethine-cyanines (3.1, $n = 1$)

$$\underset{\cdots\cdots(n+3)\pi\cdots\cdots}{}[R_2N - (CH)_n - NR_2]^{\oplus}\, X^{\ominus} \tag{3.1}$$

$$X^{\ominus} = ClO_4^{\ominus},\ Cl^{\ominus}$$
$$n = 3{,}5$$
$$R = \text{alkyl groups}$$

were published recently [131, 132]. The guanidine cation [133—141] is to be considered a branched monomethine cyanine. In a substituted pyrazoliumbromide the formation of a trimethinecyanine fragment could be evaluated [142]. The structure of dianilidopenta-methine bromide (3.1, $n = 5$, $X^{\ominus} = Br^{\ominus}$, $NR_2 = NHC_6H_5$) was determined both by G. SCHEIBE, W. HOPPE and co-workers [129, 143, 144], and by KROON and KRABBENDAM [130]. A simple tetra-cis-fixed heptamethinecyanine is the 1-methylamino-7-methyl-imino-1,3,5-cycloheptatriene (3.2).

The experimental bond lengths of these basic polymethine structures, given in Table 2, corroborate the theoretical assumptions that the bond orders p, which are correlated with the bond lengths according to well known empirical relations [146, 147], are largely

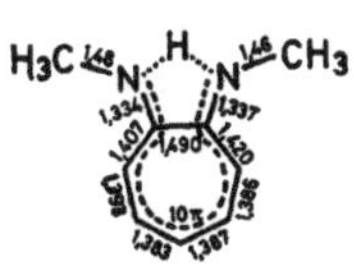

(3.2) [145]

equalised. Only in the cyclic heptamethine (3.2) and the equivalent cyclic trimethine [142] are the C⋯N and C⋯C bonds slightly lengthened at the points where the rings link together as compared to the respective mean values of Table 2, which may be ascribed to the aromatisation tendency of the ring causing a partial branching of the polymethine chain. A branching effect influences also the bond lengths in the same direction of the 1,1,3,3-tetrakis(dimethylamino)-allyl cation [148] which can be considered as a double-branched trimethine cyanine.

Whilst in the solute state the structure of the dyes was unequivocally established by

Table 2: Bond Lengths [Å] and Bond Angles In Simplest Cyanine Dyes (3.1) (Streptomethines)

NR_2	$X^\ominus$	n	Ref.	$N-C_1$	C_1-C_2	C_2-C_3	C_3-C_4	C_4-C_5
NH_2	Cl	1[a)]	[125]	1.317(7)	—	—	—	—
NH_2	Cl	1[b)]	[126]	1,314(4)	—	—	—	—
NH_2	Cl	1[c)]	[127]	1.32(1)	—	—	—	—
$N(CH_3)_2$	ClO_4	3	[118]	1.34(1)	1.39(1)	1.40(1)	—	—
$N(CH_3)_2$	ClO_4	3	[128]	1.309(9)	1.366(8)	1.395(8)	—	—
$N(C_4H_8)$	ClO_4	3	[120]	1.315(7)	1.388(8)	1.355(8)	—	—
$N(CH_3)_2$	ClO_4	5	[117]	1.312(6)	1.365(6)	1.427(6)	1.388(6)	1.399(6)
$N(CH_3)_2$	Cl	5	[123]	1.30(2)	1.38(2)	1.37(2)	1.37(2)	1.35(2)
$NH(C_6H_5)$	Br	5	[129]	1.323(5)	1.374(8)	1.376(8)	1.388(8)	1.377(8)
$NH(C_6H_5)$	Br	5	[130]	1.324(8)	1.359(8)	1.400(8)	1.381(8)	1.373(8)
Mean values of all chemically equivalent bond lengths and bond angles				1.317	1.374	1.389		
π-bond order p calculated empirically from the experimental values [147]				0.698	0.714	0.660		
HMO-π-bond order p			$n = 3$	0.647	0.647			
			$n = 5$	0.642	0.642	0.642		

a) α,α-Dithiobisformamidinium dichloride
b) Acidoformamidinium chloride
c) O-Methylisourea hydrochloride

nuclear magnetic resonance spectroscopy [94—97] and infra-red spectroscopy [98] as being completely symmetrical, there is in some dye crystals indication of a certain asymmetry of the electronic structure in the sense of a partial fixation of the valence-bond structure (3.3).

$$R_2\overline{N} - CH(=CH-CH)_m = \overset{\oplus}{N}R_2 \qquad m = 1, 2 \qquad 3.3$$

This is caused by the unsymmetrical incorporation of the anions into the crystal lattice. The trimethine structures [118, 120] as well as the pentamethine structures [117, 123, 124] crystallise in space groups which require both cations and anions to be in an unsymmetrical position. This general position results in an unsymmetrical position of the cations relative to the anions which means the two ends of the polymethine chain have different chemical environments in the crystal lattice. Consequently the structure of the cations is unsymmetrical[1]. This result, however, is within the significance points of ($\Delta m \leqq 3\sigma$), so that the conclusion to the effect of a partial fixation of a valence bond

[1] It appears, however, that this is no general rule as indicated by the structure of the bis[mono-ethylamino]-pentamethine-perchlorate now being studied [149] whose tetragonal space group $I4_1\,md$ necessitates symmetry of both the individual cation and the position of the anion relative to the cations. This is also true of the cations and anions in the crystal structure of bis-dimethyl-amino-heptamethine-chloride with the monoclinic space group C2 [150].

$C_{1(3,5)}-N$	$\sphericalangle$ $N-C_1-C_2$ ($N-C_1-N$)	$\sphericalangle$ $C_1-C_2-C_3$	$\sphericalangle$ $C_2-C_3-C_4$	$\sphericalangle$ $C_3-C_4-C_5$	$\sphericalangle$ C_4-C_5-N (C_2-C_3-N)
1.29	123.0°	—	—	—	—
1.302	123.1°	—	—	—	—
1.29	123.8°	—	—	—	—
1.32(1)	123.1°	116.2°	—	—	125.2°
1.306(8)	126.3°	118.0°	—	—	120.1°
1.319(7)	126.2°	117.4°	—	—	126.9°
1.323(6)	124.8°	117.8°	122.6°	118.6°	126.6°
1.38(2)	125°	118°	125°	118°	128°
1.326(5)	124°	122.5°	125.2°	123.4°	123.4°
1.332(8)	123°	123.2°	124.6°	121.2°	123.3°
	124.2°	119.0°	124.3°		

structure can only be a tentative one. In the trimethineoxonols (Section 3.2) the effect seems to be significant.

Unequivocally established have been the deviations from the 120° bond angle, typical in aromatics for sp^2 hybridised carbon atoms [64]. In the simple polymethine chains, the carbon-carbon valence angles at centres with a high π-electron density (negatively charged atoms) are markedly smaller than 120° and greater than 120° at the centres with a low π-electron density (positively charged atoms): Table 2. These effects are obviously attributed to the rehybridisation effect dependent on the total electron density at the atoms Q which may be described approximately by the simple relationships (1) and (2) [64]:

$$\lambda^2 = 2 - Q \tag{1}$$

$$\cos \theta_{ij} = \frac{1}{Q - 2} \tag{2}$$

where λ = hybridisation parameter and θ_{ij} = C—C—C and N—C—C bond angles within the polymet hine chain. The effect was subsequently established by theory based on the principle of maximum overlap [151], giving results only qualitatively in agreement with the experimental data.

A further indication of electron density alternation in the polymethines is provided by geometrical peculiarities in the intermolecular arrangement of polymethine-like molecules, which have been detected independently by KULPE [152] and SCHEIBE [153]. In the cases studied so far, typical dye molecules orient relative to each other in the crystal lattice in such a way that the positively charged centres in the polymethine chain of a molecule hold the shortest possible distance to the negatively charged centres of the adjacent molecule. It should however be pointed out in this context that, by definition, charge density alternation is valid for the atomic centres. When calculating the electric potential at the periphery of a polymethine molecule from the quantum-chemically determined π- and σ-electronic charges, there is no evidence of a charge alternation outside the VAN DER WAALS' radius of the polymethine chain [154, 155].

Intermolecular packing of organic molecules is a question of great relevance to the aggregation characteristics of organic dyes. As is well known, such aggregates are capable of energy transfer and play a significant role in the spectral sensitisation of photographic emulsions [102, 156—159]. There is a possibility that the energy transfer mechanism in the spectral sensitisation of photographic silver halide emulsions is governed by epitaxial relationships between aggregates and matrices. As a result of aggregation in dye solutions, in adsorbates and in crystals, new absorption and luminescent bands may appear which are displaced in the direction of longer and/or shorter wavelengths compared to the long wavelength $S_0 \rightarrow S_1$ transition of monomeric molecules. To explain this phenomenon, various hypothetical models of sandwich, brick-stone-work, step-like or ladder structures of associated dye molecules have been proposed [160—169]. In recent time first aggregate structures have been resolved by X-ray analysis [102, 156, 170].

A thread-like arrangement of molecules is observed in the crystals of pseudoisocyanine [153]. Also arranged in the form of long threads running through the crystal lattice, are the molecules of the bis-[dimethylamino]-pentamethine cation both in orange coloured crystals of the perchlorate [117, 119] and in the deep-violet crystals of the chloride [123]. The two states of aggregation differ however considerably in regard of the angle of dislocation between two molecules and in the arrangement of the threads in the crystals relative to each other [102].

Contrary to SHEPPARD's conclusion [171], water molecules contained, for instance, in the crystals of bis-[dimethylamino]-pentamethine chloride [123], do not maintain the union of the dye cations inside the aggregated threads by coordination, rather they shield the chlorine anions from the dye cations by Cl···H—O hydrogen bridges, this way facilitating the hydrophobic interaction among the cations [102, 156]. The same applies to the structures of other, crystal-water containing dyes, such as those of methylene blue [172]. An excellent survey covering the aggregation in cyanine dyes as disclosed by diffraction techniques is given by D. L. SMITH [156, 170]. These studies demonstrate the fundamental role played by X-ray diffraction methods in clarifying principal questions of dye theory with regard to industrial-technical applications.

A number of X-ray analyses of more complicated polymethinecyanine dyes have been carried out. Trimethine fragments are contained in the pseudoisocyanine chloride (3.4), 3,3'-diethyl-thiacyanine bromide (3.5), and the 1,1',3,3'-tetraethylimidazolo-[4,5-b]-quinoxalino-cyanine iodide (3.6). Less accurate data for the bromide of the pseudoisocyanine [176] and of a $(TCNQ)_2$-complex of the 3,3′-dimethyl-benzthiacyanine bromide [177] have also been reported. Crystal analysis data exist on the unsymmetrical N,N′-dimethyl-quina-(2)thia-(2′)-cyanine iodide [178].

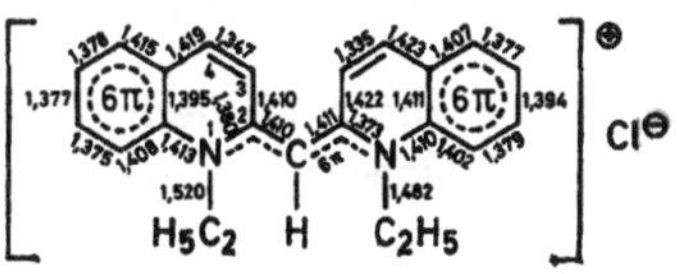

(3.4) [173]

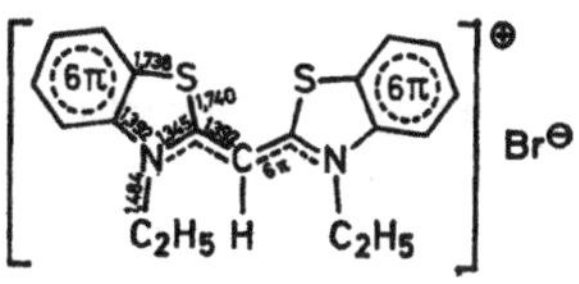

(3.5) [174]

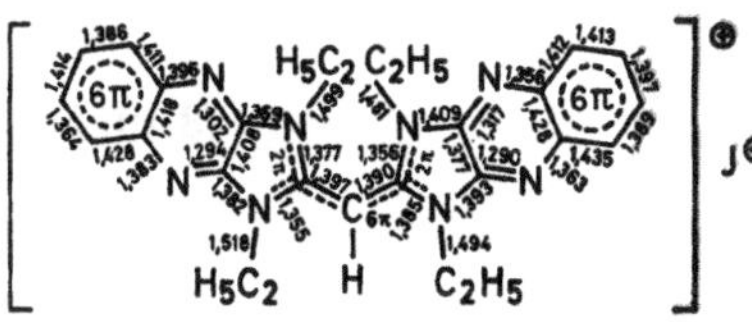

(3.6) [175]

ROMERS et al. [179, 180] investigated by means of *X*-ray structure analysis the tautomerism of 2,2-methylene diquinoline (3.7) discovered by SCHEIBE et al. [181, 182]. To date, accurate interatomic distance have been obtained only for the colourless form, in which the polymethinic conjugation is blocked by a methylene group. The red form exists in a di-cis configuration [180].

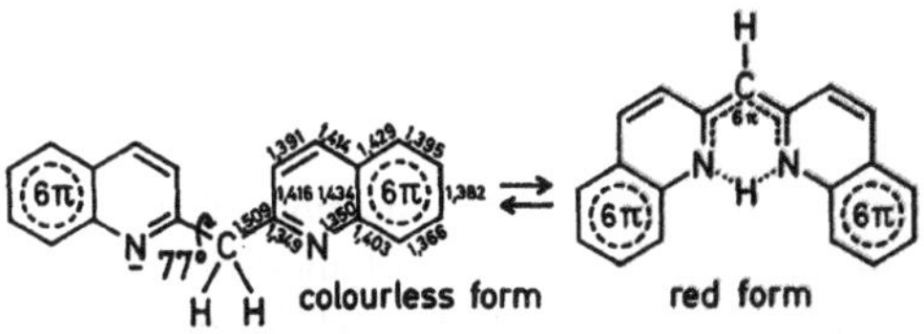

(3.7) [179]

Pentamethine structures are realised in the N,N′-diethyl-thiacarbocyanine (3.8) and its meso-substituted derivative [183], in the N,N′-diethyl-2,2′-thiazolinocarbocyanine iodide [184] and in two solvates of 5,5′,6,6′-tetrachloro-1,1′,3,3′-tetraethyl benzimid-azolocarbocyanine iodide (3.9). Recently the structure of the latter dye molecule with a triple bond within the pentamethine chain was determined [186].

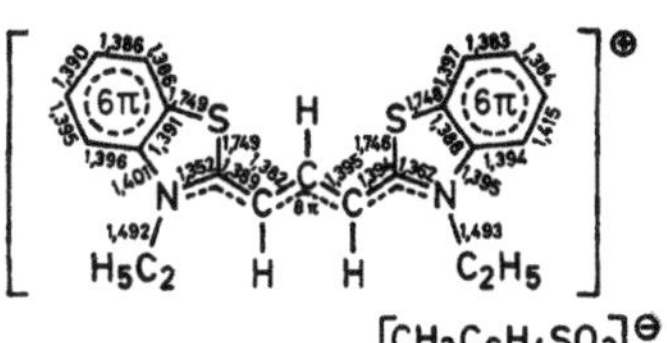

(3.8) [124]

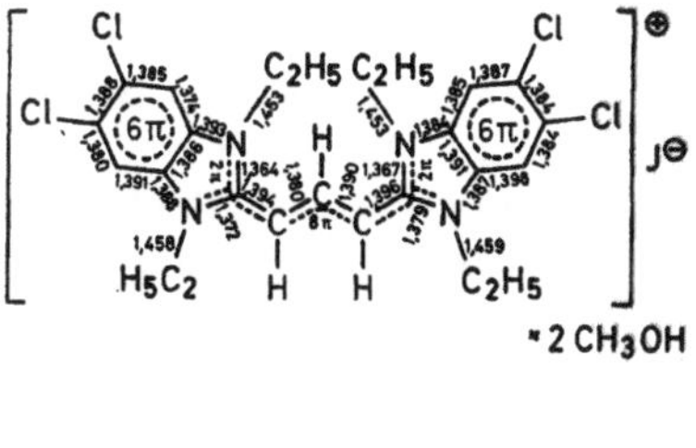

(3.9) [185]

The structures of thiacarbocyanine bromide and its ethanole solvate were established as early as in 1959 [187, 188]. In the meantime X-ray analyses of the same dye cation have been made, using chloride anion [189], aurate anion [190] and para-toluene sulfonate anion [124, 191], of which the latter yields the most accurate data. These are presented in formula (3.8).

A heptamethinecyanine structural group is to be found in the N,N'-diethyl-4,4'-quinacyanine bromide (3.10), and repeatedly branched nonamethinecyanine fragments in the triphenylmethane dye pararosaniline (3.11).

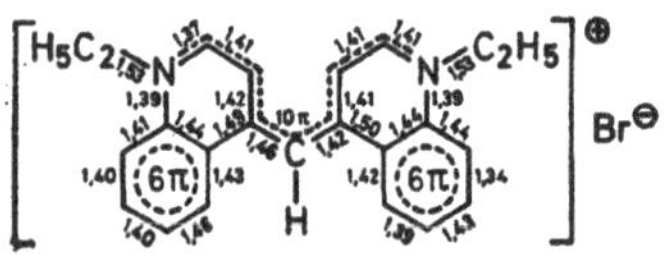

(3.10) [176]

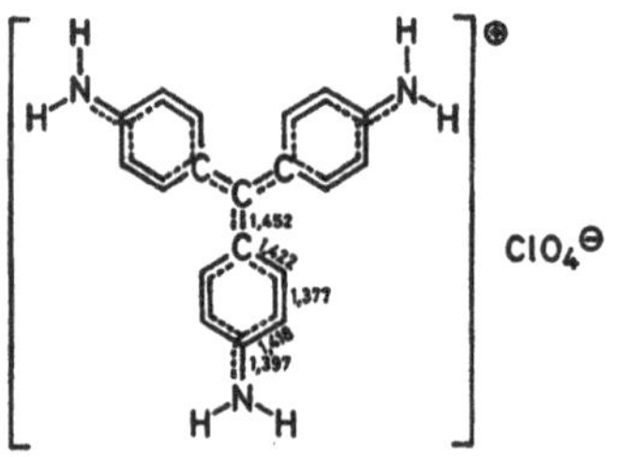

(3.11) [192]

A common feature of all structures is the high degree of equalisation of the C=C bond lengths as well as of the N=C bond lengths within the polymethine chain A comparison with unbranched chain molecules (3.1) reveals a distinct lengthening of the C—N bond, from about 1.32 Å to about 1.37 Å, owing to chain branching. From a comparison with the tetrazolium ion (3.12) and other tetrazole derivatives [194, 195] with C=N bond lengths of about 1.32 Å, similar to those in the open-chain cyanines, it follows that the bond extension is indeed caused by a branching effect and not by the effect of cyclisation. This is confirmed by the structure of (3.13) where branching of the aromatic resonance also gives rise to the expected extension. An exception to this rule appears to be the simple benzimidazoles, where, in some cases, C=N bond lengths of 1.35 Å have been observed in the ring [197—199].

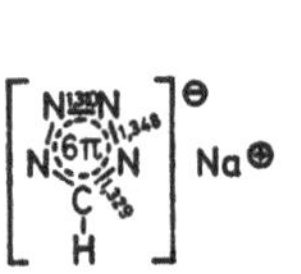

(3.12) [193]

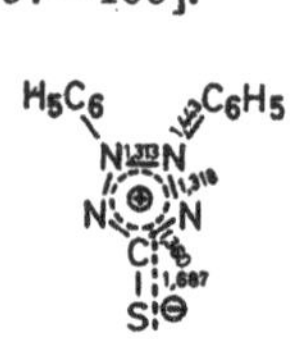

(3.13) [196]

In the condensed benzene rings of dyes (3.4) to (3.9) an aromatic structure with an average bond length of about 1.4 Å is realised to a large extent. Of interest here is the pronounced double-bond character of the $C_3—C_4$ bond with an average length of 1.34 Å in the quinoline ring of dye (3.4), resulting from the preferential realisation of the trimethine chain and the two aromatic 6π ring systems. For comparison, the same bond length in an 8-hydroxyquinoline complex is 1.40 Å [200]. A similar double-bond fixation is also found in the interior rings of the dye (3.6). Dominating here are obviously the aromatic 6π systems of the outer rings and the branched trimethine system. In the 4,4′-quinacyanine (3.10), in contradiction to the 2,2′-linked quinacyanine (3.4), there occurs no double-bond fixation. Here the heptamethinecyanine structure is obviously dominating, which induces a lengthening both of the C—C bonds joining the benzene ring and of the adjacent aromatic bonds of the condensed benzene ring.

Another consequence of branching polymethinic resonance systems is to be found in the pararosaniline (3.11), whose benzene rings show deviations in the direction of a slightly polyenic behaviour as the symmetric branching of the nonamethine system at the para positions of the rings gives rise to an extension of the adjacent bonds within the rings. In consequence of this enlargement the two central bonds of the rings are shortened. The stronger the polymethinic resonance the more pronounced is this effect as can be seen by comparison with methylene blue (this Section, (3.24)) or WURSTER Blue radicals (Section 7.1).

For a detailed discussion of the thiacyanine structures (3.5) and (3.8) a comparison with the structures of the thiazolium ion (3.14) and benzothiazolium ion [202] as well as of a variety of benzothiazole derivatives [203—209], thiadiazoles [210, 211] and thiazolinones and related compounds [212—219] is of interest.

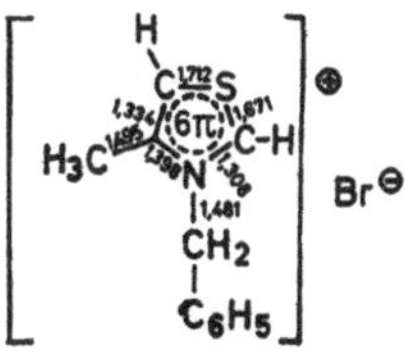

(3.14) [201]

Formulae (3.15) and (3.16) are examples representing the anhydrous base of the 2-methylbenzothiazole and the 2-methylaminobenzothiazole. Amino-substituted thiazolinones contain an aza-trimethine-merocyanine group, hence they have been classified under Section 4.3.

A C—S single bond length of ≧ 1.81 Å (cf. Table 1) is to be observed in the left benzothiazole ring of (3.15). All remaining C—S bonds show a more or less equalised bond length between that of a single and a double bond, in consequence of the formation

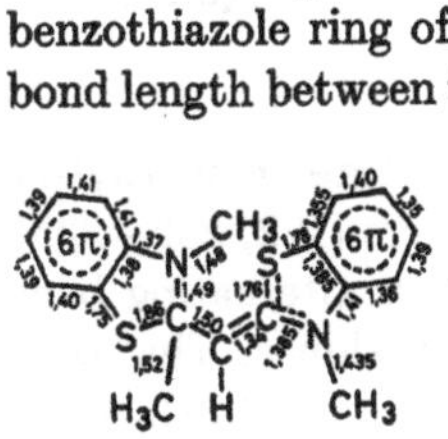

(3.15) [206]

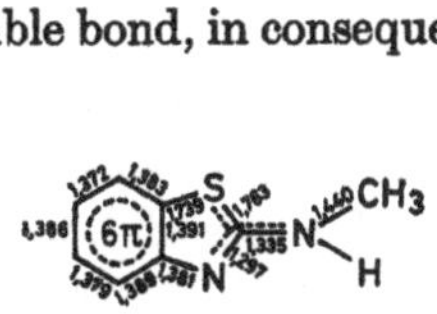

(3.16) [207]

of an aromatic structure as in (3.14) or of the existence of monomethine-like structures as in (3.15) and (3.16). On the average, C⩵S bond lengths between 1.70 and 1.75 Å are to be found, whilst the C⩵N bond lengths range between 1.30 and 1.40 Å.

In this connection reference is to be made to the particular, not yet conclusively explained, electronic structure of the 6a-thiathiophthenes and their heteroatom analogous derivatives which are reviewed in [110] and [220b]. Obviously the colour of these substances is likewise the result of a strong delocalisation of the π-electron system with an alternating electron density distribution, involving resonance structures (3.17) to (3.19). The polymethinic character of structures (3.17) and (3.19) with 8π-electrons on 7 atomic centres is evident.

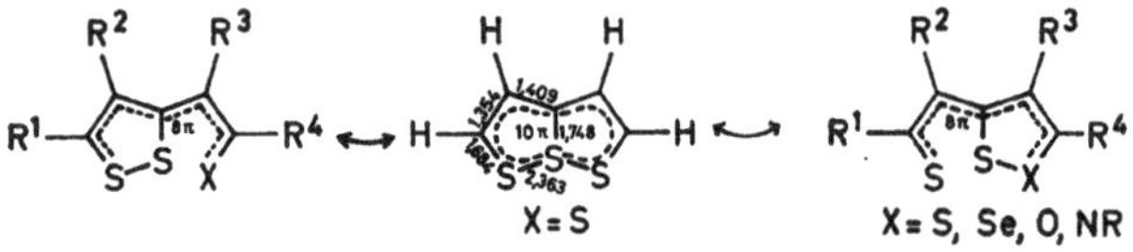

(3.17) (3.18) [220a] (3.19).

The bis-[N-ethyl-benzothiazole-(2)]-phosphamethinecyanine perchlorate (3.20) represents a symmetrical trimethinecyanine with heteroatom substitution, constituted analogously to the thiacyanine (3.5). Corresponding with the polymethinic resonance over the phosphorous atom, the P═C distance of about 1.76 Å is markedly shorter than the 1.83 Å of the triphenyl phosphine [222]. The structure of the bis-(1-ethyl-2-quinolyl)-phosphamethinecyanine perchlorate has also been reported [223]. As expected it coincides largely with the analogous quinacyanine (3.4).

A trimethine chain substituted by a nitrogen heteroatom is contained in the dimethyl-[3-dimethylamino-2-aza-propene-(2)-ylidene]-ammonium-tri-iodide (3.21), the morpholine-

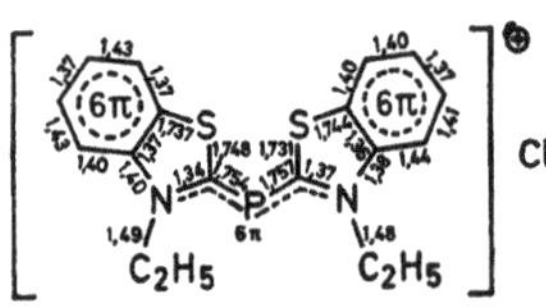

(3.20) [221]

diguanide-hydrobromide [225], and the thiuret, e.g. the 3,5-diamino-1,2,4-dithiazolium ion (3.22) [226—228]. A branched tetraaza-pentamethine structure is to be found in the triazido-carbonium-hexachloroantimonate [229], aza-heptamethine structures in the proflavine cation [230] and in the bis-10-methylacridine-9-monoaza-mono-

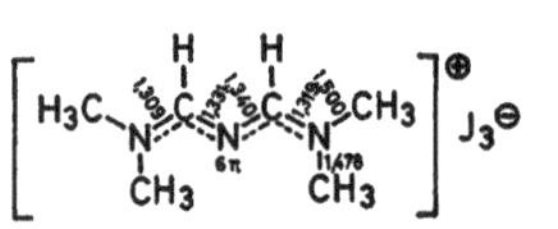

(3.21) [224]

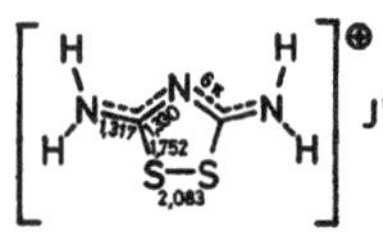

(3.22) [226]

methinecyanine perchlorate (3.23), and an aza-nonamethine structure in methylene blue (3.24), whose structure has been determined by several workers [172, 232—235].

With their four C═N bonds of 1.32 Å, on the average, and their alternating bond angles, the simple aza-trimethinecyanines are nearly ideal polymethines. The azido salt [229] resembles in its structure the cyanopolymethines (3.26) and (3.27), which will be discussed later.

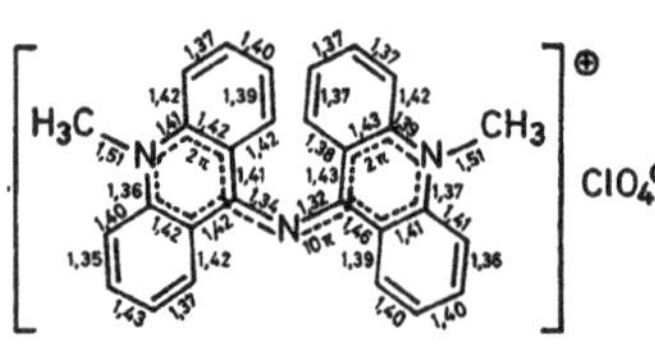

(3.23) [231]

The acridine dye cation (3.23) and some of its derivatives [236, 237], too, with its repeatedly branched heptamethine structural group and four aromatic 6π rings, display such bond lengths as would be expected from the polymethine concept point of view. In methylene blue (3.24), on the other hand, there are indications of polyene-like

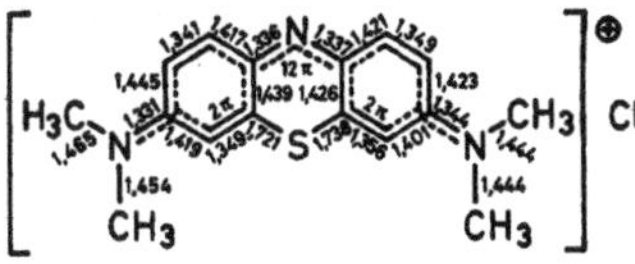

(3.24) [172]

structures, with alternating double and single bonds. As in (3.11) this seems to be the consequence of branching the polymethinic resonance by the carbocyclic systems. In addition, the two sulphur-carbon bonds with 1.73 Å are shorter than the C—S single bond length (Table 1) which indicates an additional heptamethine resonance involving the sulphur atom. Also in the unsubstituted parent compounds the C—S bond lengths are on the average not smaller than 1.76 Å [238—241]. The bond lengths of (3.24) are very similar to the structure of WURSTER's dyes which will be discussed in conjunction with the polymethine radicals (cf. Section 7.1).

The methylene blue structure is of great interest in connection with the intermolecular aggregation effects which are not explicitly dealt with herein. Sandwich-like pairs of molecules have been detected in the crystal lattice [233] which could explain the occurrence of aggregate absorption bands at shorter wavelengths. If water molecules are present in the crystal lattice, these will form hydrogen bridges with the anions [172], thereby shielding the cation aggregates from the anionic charges as in the simple polymethinecyanines mentioned before [102]. Water is a necessary condition for the blue colour of the dye which is red-coloured in water-free solutions [242, 243].

Other than the above discussed cyanine dyes with a cationic character, the dye (3.25) contains an anionic trimethinecyanine structural group. The branching of the

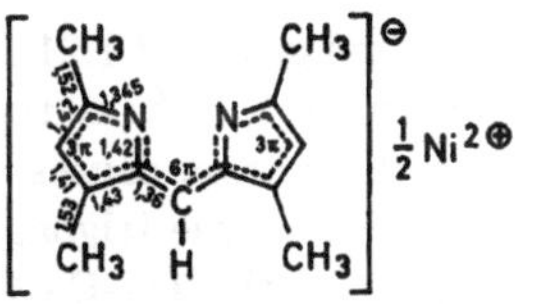

(3.25) [244]

conjugated system in the rings is apparent. In this case a particularly strong double bond fixation occurs at the two central C=C methine bonds (1.36 Å). A comparable but azasubstituted anionic trimethinecyanine fragment is to be found in the (2,2'-dipyridyliminato)palladium(II) complex [245], whose ligand in its neutral form has also been investigated for comparison [246, 247].

Anionic polymethine structures with nitrogen terminal atoms are further realised in the alkali salts of the cyanopolymethines [248—252], of which examples are given in (3.26) and (3.27).

In the nonamethine chain of (3.27) there is indication for the expected bond length equalisation, while again lengthening of the bonds occurs at the branching atoms. Surprisingly the C≡N bond distance is not extended compared to the non-conjugated cyano group with an average 1.15 Å (Table 1) [253, 254]. There obviously exists a certain analogy to the carbonyl group where a lengthening is likewise difficult to attain; the effect will be described in detail in Sections 4.2 and 6.2.

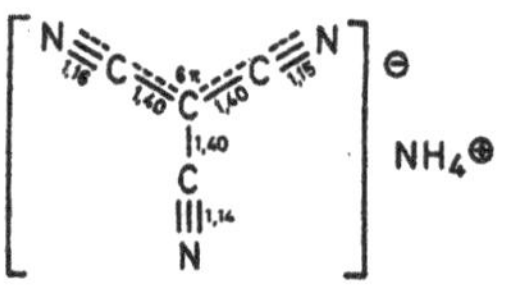

(3.26) [251]

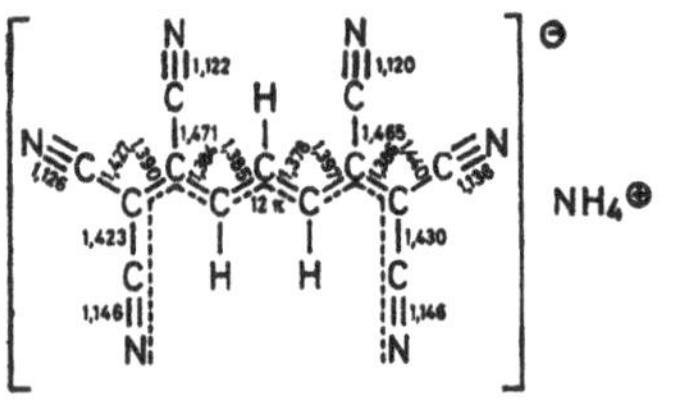

(3.27) [252]

3.2. Oxonols

According to the polymethine concept, a polymethine-like structure should be realised also in the symmetrical oxonol dyes with oxygen atoms as terminal groups (2.1, $Y = CR$; $X = X' = O^{1/2\ominus}$).

The carboxylate anions (3.28) contain monomethine structural groups. The X-ray structures of these anions have been critically reviewed by several workers [68, 146, 255a, 255b].

Average bond lengths taken from Table 3 (3.28)

Equalisation of the bond lengths of a carbonyl group and a hydroxyl group of 1.21 Å and 1.37 Å, respectviely, (Table 1) yields an average bond length of 1.26 Å in the carboxylate ions (Table 3). The high electron deficiency on the methine atom causes a rehybridisation according to Equations (1) and (2) (Section 3.1) from the true trigonal state towards the digonal state and consequently, causes the angle of the O—C—O bond of 120° to be enlarged to between 124° and 126°. By replacing the substituent R on the methine atom by a strong electron acceptor, e.g. by a trifluoromethyl group [262], a carbamide grouping [263] or a carbonyl group [264], this angle will be enlarged to as much as 128°. Conversely, the angle will be reduced to 123° when the substituent is an electron donor as in the carbamates [265, 266].

Trimethineoxonol structures are realised in the metal chelates of acetylacetone (3.29), on which extensive experimental evidence is available [267—270]. By contrast to the simple trimethinecyanines which are present in an all-trans structure, the tri-

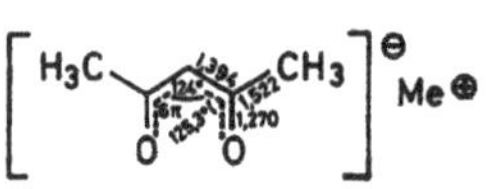

Average bond lengths and angles taken from Table 4 (3.29)

Table 3: Bond Lengths [Å] and Angles In Carboxylate Ions (3.28)

No.	Substance	*Me*	Ref.	$C-O_1$	$C-O_2$	$C-R$	∢ O—C—O
1	Formate ion (Averages)	Na Ca Sr Ba	[70]	1.25	1.25	—	125.5°
2	Sodium formate	Na	[256]	1.246	1.246	—	126.3°
3	Sodium hydrogen diacetate	Na	[257]	1.295	1.243	1.492	121.6°
4	Lithium glycolate	Li	[258]	1.255	1.254	1.525	125.0°
5	Ammonium oxalate monohydrate	NH_4	[259a]	1.263	1.252	1.569	126.0°
6	Lithium oxalate	Li	[259b]	1.264	1.252	1.561	127.3°
7	Disodium maleate	Na	[260]	1.265	1.280	1.500	125.6°
8	Sodium pyruvate	Na	[261]	1.278	1.242	1.579	126.3°
9	Ammonium oxamate	NH_4	[262]	1.256	1.248	1.564	128.1°
10	Ammonium trifluoroacetate	NH_4	[263]	1.274	1.264	1.542	128.3°
Averages				1.265	1.253	1.543	126.0°

methineoxonols exhibit a di-cis configuration due to chelating. This applies also to the salts with alkali or alkaline earth ions [271, 272]. The bond lengths obtained from some later and more precise *X*-ray structures have been compiled in Table 4.

Here too the bond lengths are largely equalised throughout the methine chain, including a lengthening of the carbonyl group, so that these cyclic compounds had until recently been assigned a quasi-aromatic structure [267, 290, 291]. When it was established by nuclear magnetic resonance spectroscopic studies that it was impossible to induce in the magnetic field a ring current characteristic for aromatics [4, 5, 292, 293], the concept of aromaticity was dropped without an alternative being advanced at first [2]. With the triad system [6—9] it has become possible to classify the acetyl acetonates as polymethineoxonols and to explain in that way their long wavelength absorption band, their tendency to electrophilic and nucleophilic substitution as well as their bond length equalisation in the absence of a ring current indicating aromaticity.

The cyclic di-cis structure in the chelates makes it difficult to interpret the bond angles with respect to rehybridisation effects caused by alternating electrondensities. Nevertheless, in analogy with open-chain polymethines, a somewhat larger angle of about 125° is observed on centres with a low π-electron density (O⩵C⩵C angle) and a somewhat smaller angle on the central atom with a high π-electron density (Table 4). By substituting trifluoromethyl groups with strong electron acceptor properties for the methyl groups in (3.29), the C⩵C⩵O angle is caused to enlarge significantly from 125° to about 130° [294—296]. This increase, which the authors were unable to explain, can be conveniently interpreted by the polymethine concept, in accord with Eq. (2).

In a number of chelates an unsymmetrical molecular structure is realised in spite of the symmetrical constitution of the molecule which is obviously attributable to specific crystal lattice effects. These compounds will be discussed in Section 4.3. under Merocyanines.

Tabel 4: Bond Lengths [Å] and Angles In Acetyl Acetonates (3.29)

No.	Substance	*Me*	Ref.	C—O*)	∢C—C—O*)	C—C*)	∢C—C—C	$C—CH_3$*)
1	Acetylacetonato-lithium	Li	[271]	1.260	123.2°	1.401	126.7°	1.495
2	Bis-acetylacetonato-beryllium(II)	Be	[272]	1.26	124°	1.39	121°	1.50
3	Tris-acetylacetonato-aluminium(III)	Al	[273]	1.269	125.1°	1.393	121.4°	1.535
4	Tris-acetylacetonato-gallium(III)	Ga	[274]	1.261	125.2°	1.390	124.3°	1.525
5	Bis-acetylacetonato-4-aminopyridine-copper(II)	Cu	[275]	1.254	126.3°	1.398	123.9°	1.530
6	Diaquotris-acetylacetonato-lanthanum(III)	La	[276]	1.290	123.5°	1.410	125.7°	1.532
7	Tris-acetylacetonato-chromium(III)	Cr	[277]	1.263	125°	1.388	125°	1.517
8	Di-μ-diphenylphosphinato-acetylacetonato-chromium(III)	Cr	[278]	1.268	125.8°	1.395	123.2°	1.519
9	Tris-(1,3-propanedionato)chromium(III)	Cr	[279]	1.283	125,7°	1.385	124.5°	—
10	Tris-acetylacetonato-iron(III)	Fe	[280]	1.258	125.0°	1.377	124.8°	1.530
11	Hexa-acetylacetonato-aquo-tricobalt(II)	Co	[281]	1.268	125.7°	1.391	125.6°	1.512
12	Tris-acetylacetonato-cobalt(III)	Co	[282]	1.267	125.4°	1.382	125.5°	1.494
13	Bis-acetylacetonato-bis-(pyridine-N-oxide)nickel(II)	Ni	[283]	1.270	125.6°	1.374	126.9°	1.519
14	Tris-acetylacetonato-ruthenium(III)	Ru	[284]	1.27	126°	1.40	126°	1.53
15	Tris-[1,1,1,5,5,5-hexamethyl-acetylacetonato-bispyridine-europium(III)	Eu	[285]	1.257	123.9°	1.390	124.9°	1.512
16	Tris-acetylacetonato-ytterbium(III)	Yb	[286]	1.288	125.2°	1.422	121.8°	1.544
17	Tris-acetylacetonato-aquo-ytterbium-(III)-hemibenzene	Yb	[287]	1.285	125.0°	1.403	122.6°	1.552
18	Tris-acetylacetonato-(4-amino-3-pentene-2-one)-ytterbium	Yb	[288]	1.29	123.9°	1.41	124°	1.54
	Averages			1.270	125.0°	1.394	124.2°	1.522
	Averages of 29 independent data from 13 three-dimensional *X*-ray structure analyses		[289]	1.274	125.3°	1.390	124.02°	1.524

*) Averages of mirror- symmetrically equivalent bonds

The behaviour of the trimethineoxonol chain on substitution of additional, conjugated substituents, e.g. of phenyl groups [297, 298] is shown in (3.30). The planes of the phenyl groups are twisted through a dihedral angle of only 9.3° and 4.6°, respec-

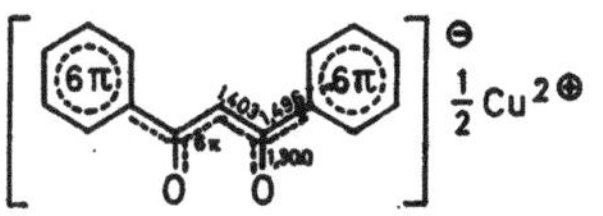

(3.30) [297]

tively, relative to the planes of the trimethine ion. This enables the phenyl groups to contribute to the resonance system of the trimethine. Consequently the bonds involved in the branching of the π-electron system are caused to enlarge which, in this case, affects in particular the carbonyl bonds.

Branched diaza-trimethineoxonol fragments include the alkali salts of the trinitromethane (3.31, $R = NO_2$) [299] and the 2,4,6-trinitrophenyldinitromethane (3.31

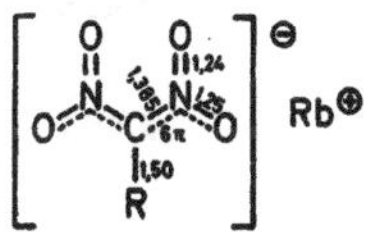

R = NO₂

(3.31) [300]

$R = 2,4,6\text{-}(NO_2)_3C_6H_2$). The bond lengths of the latter are given in (3.31). The interatomic distances within the trimethine group are strongly equalised. To compare, the average bond lengths of the nitro groups on the phenyl ring are 1.51 Å for C—N, but only 1.21 Å for N=O.

The deeply coloured complexes (3.32) [301, 302] first synthesised by Meisenheim can be included in the same category of heteroatom substituted oxonols. Shortening of the C—N bonds and lengthening of the N—O bonds is more pronounced in the central nitro group than in the two outer nitro groups. This points to a branched pentamethine structure, according to formula (3.32). In principle, the formation of a diaza-heptamethine structure between the two outer nitro groups would have been possible. A similar effect is to be observed in (3.33) and in the potassium salt of 4-hydroxy-5,7-dinitrobenzofuroxane [304], where the one-sided preference of resonance within a diazapentamethine structure leads to an even more pronounced bond length equalisation.

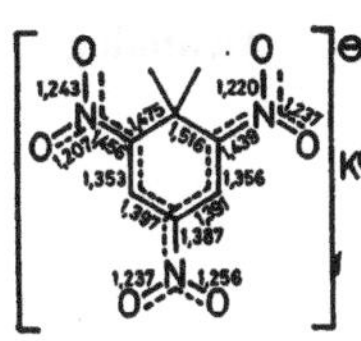

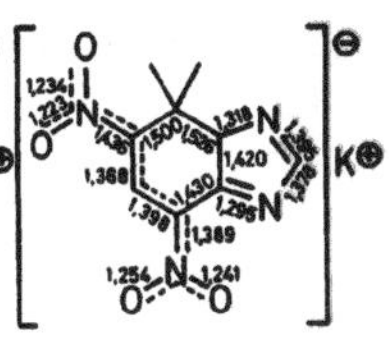

(3.32) [301] (3.33) [303]

Apparently caused by the multiple branching of the mesomeric system, the degree of deviation from the ideal bond length equalisation is somewhat greater in the complexes of Meisenheim. The interatomic spacings for true polyenes, however, are not reached. The reference values of the carbon-carbon bond lengths are 1.33 and 1.47 Å, respectively, and of an unconjugated nitro group 1.23 Å, for N=O, and 1.47 Å, for C—N, (Table 1). These values are, on the average, actually attained in the non-ionic 4,6-dinitrobenzofuroxane [305].

While simple pentamethineoxonol structures have not been studied thus far, a tetra-cis-fixed heptamethineoxonol structural group (3.34) was found to be present in the tropolonate ion, whose bond lengths are summarised in Table 5.

Table 5: Bond Lengths [Å] In Tropolonate Ions (3.34)

No.	Substance	Me^+	Ref.	C_1-O (C_7-O)	C_1-C_2 (C_6-C_7)	C_2-C_3 (C_5-C_6)	C_3-C_4 (C_4-C_5)	C_1-C_7
1	Sodium tropolonate	Na	[306]	1.278	1.426	1.390	1.388	1.487
2	Tris(tropolonato)-aluminium(III)	Al	[307]	1.291	1.397	1.384	1.371	1.450
3	Tris(tropolonato)-monochlorotin(IV)	SnCl	[308]	1.287	1.412	1.392	1.393	1.453
4	Tris(tropolonato)-monohydroxotin(IV)	SnOH	[308]	1.285	1.396	1.377	1.380	1.458
5	Tris(tropolonato)-iron(III)	Fe	[309]	1.294	1.397	1.385	1.379	1.463
6	Di-μ-tropolonato-bis-[aquo(tropolonato)-nickel(II)]	Ni	[310]	1.294 (1.264)	1.402 (1.422)	1.415 (1.398)	1.380 (1.371)	1.486
7	Tetrakis-(tropolonato)-N,N'-dimethylformamide-thorium(IV)	Th	[311]	1.277	1.407	1.385	1.383	1.471
	Averages			1.285	1.406	1.385	1.382	1.464
	Reference values	Formula		$C-N$	C_1-C_2 (C_6-C_7)	C_2-C_3 (C_5-C_6)	C_3-C_4 (C_4-C_5)	C_1-C_7
	N-analogue	3.2	[145]	1.336	1.414	1.390	1.385	1.490
				$C-S$				
	S-analogue	3.39	[312]	1.720	1.397	1.389	1.382	1.448

The analogous nitrogen compound (3.2) was mentioned under Cyanines. An analogous sulphur compound (3.39) will be discussed in Section 3.3.

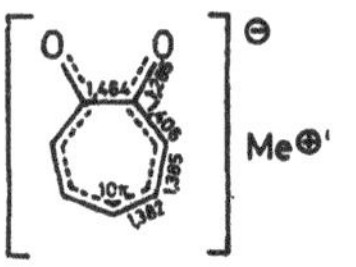

Average bond lengths taken from Table 5 (3.34)

The C⎓O spacings are lengthened to a greater extent than in other typical oxonols with 1.26 to 1.28 Å. The C⎓C distances are nearly identical with those of the corresponding cyanines (3.2, Table 2). There again the bonds involved in the ring-branching are somewhat enlarged, compared to the corresponding bond lengths in the all-trans cyanines.

A diaza-heptamethine structural group is present in the potassium dinitro-trichloro-cyclopentadienide (3.35).

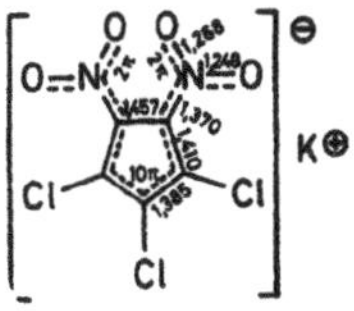

(3.35) [313]

Its structure confirms well the points just outlined, the N=O bonds are enlarged, the bond lengths within the heptamethine chain are equalised and the branching C_1-C_2 ring bond is relatively long. Diaza-heptamethineoxonol-fragments are also realised in the 3,5-diacetamino-1,2-dithiolium bromide [314] and probably more strengthened in its anionic form, the 3,5-diacetylimino-1,2-dithiole trihydrate [315], whose structure analysis is not very accurate.

3.3. Other Terminal Atoms on Polymethine Chains

Symmetrical monomethine structures with sulphur terminal atoms are realised in the dithiocarboxylate ions, of which the structures of tetrakis-(dithioacetato)-vanadium-(IV) [316] and bis-(dithiobenzoato)zinc-(II) [317] have been reported. The thoroughly studied dithiocarbamates (3.36) also contain monomethinethionol structures. The latter are summarised in Table 6.

Table 6: Bond Lengths [Å] In Dithiocarbamates (3.36)

No.	Substituent *R*, *R*	*Me*+	Ref.	C—S	C—N
1	CH_3, CH_3	Sn^{IV}	[318]	1.735	1.31
2	CH_3, CH_3	Cu^{II}	[319]	1.721	1.31
3	CH_3, CH_3	Zn^{II}	[320]	1.718	1.34
4	CH_3, CH_3	Zn^{II}	[321]	1.722	1.346
5	CH_3, CH_3	Fe^{III}	[322]	1.726	1.317
6	CH_3, H	Ni^{II}	[323]	1.71	1.30
7	$(C_2H_5)_2$	Sn^{II}	[324]	1.712	1.358
8	$(C_2H_5)_2$	Sn^{IV}	[325]	1.72	1.31
9	$(C_2H_5)_2$	Pb^{II}	[326]	1.703	1.37
10	$(C_2H_5)_2$	Te^{II}	[327]	1.72	1.34
11	$(C_2H_5)_2$	Cu^{II}	[328]	1.717	1.34
12	$(C_2H_5)_2$	Zn^{II}	[328]	1.727	1.32
13	$(C_2H_5)_2$	Cd^{II}	[329]	1.720	1.33
14	$(C_2H_5)_2$	Mn^{III}	[330]	1.71	1.36
15	$(C_2H_5)_2$	Co^{III}	[331]	1.704	1.32
16	$(C_2H_5)_2$	Ni^{II}	[328]	1.706	1.33
17	$-(CH_2)_6-$	Zn^{II}	[332]	1.740	1.315
18	$(nC_3H_7)_2$	Cu^{II}	[333]	1.721	1.33
	Average values			1.719	1.330

The C—S distances of 1.72 Å for sulphur-containing monomethines are clearly intermediate between those of the single and the double bond lengths (1.56 and 1.81 Å, respectively, Table 1). The C—N distance of 1.33 Å in (3.36) is evidence of a relatively

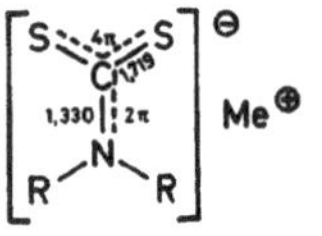

Average bond lengths taken from Table 6 (3.36)

strong additional resonance of the amino groups and hence, of a branching of the conjugated system. This branching effect is to be expected to be even stronger in xanthato or thioxanthato complexes, i.e. by replacing the amino group by a methoxy or alkylmercapto group, owing to the observed enlargement of the S—C—S angle [334].

A trimethine structural group with sulphur terminal atoms is formally located in the bis[dithioacetylacetonato]-tetrachloroferrate-(II) [335]. Here more recent structure analyses have proved the molecule to possess an intramolecular S—S bond, so that this compound and its derivatives [336, 337] must be regarded as dithiolium complexes of iron (3.37) which can be compared in their structure with other dithiolium ions [338—342]. Especially in the 3,5-diamino-1,2-dithiolium cation (3.38) and, as mentioned before, in the thiuret ions (3.22) it is possible to formulate in addition a trimethinecyanine

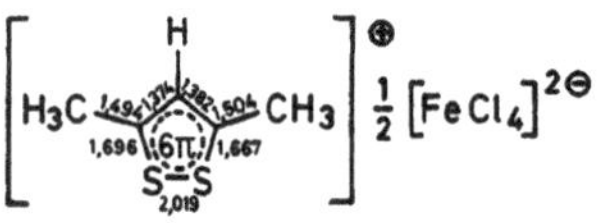

(3.37) [335]

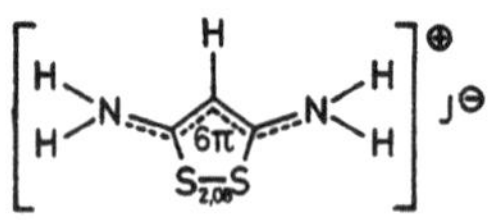

(3.38) [343]

fragment. Hordvik already showed [343] that the S—S and the C—S bond lengths in the ring are the longer, that is the aromatic character of the dithiolium ring is the weaker, the more pronounced is the polymethine resonance within the trimethine chain. The 3,5-diacetylamino-1,2-dithiols [314, 315] mentioned in the preceding section under Azaheptamethineoxonols will have to be considered in the same context.

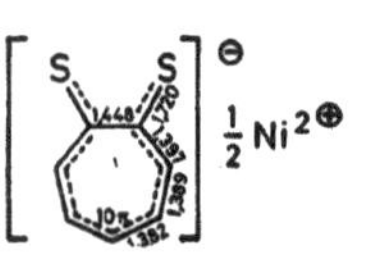

(3.39) [312]

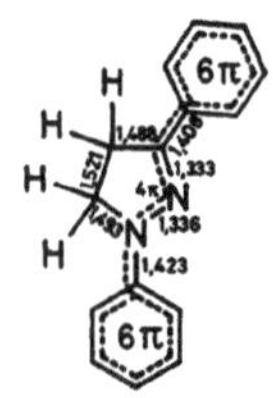

(3.40) [344]

A heptamethine structure with sulphur terminal atoms is present in the bis-(dithiotropolonato)-nickel-(II) (3.39) whose bond lengths coincide with the expected values, as follows from a comparison with the respective cyanine (3.2) or oxonol (3.34).

The structures of simple symmetrical polymethines with terminal carbon atoms (to be found, for example, in many carbenium ions [74]) have not been determined so far. There is a high degree of bond length equalisation in the diphenyl-Δ^2-pyrazoline (3.40), suggesting the presence of a diaza-trimethine structural group with carbon terminal atoms and a largely symmetrical π-electron distribution.

4. Polymethine-Polyene Intermediates

4.1. Principles

Intermediates between ideal polymethines (2.1) and polyenes (1.2) result when the electronic structure of the polymethines becomes increasingly unsymmetrical as in Formulae (2.3) and (2.4), owing to the different electronegativity of the X, X' chain terminal atoms. Typical representatives of this group are the simple merocyanines, the electronic structure of which may be varied as a function of the electronegativity of the NR_2-substituents as well as by environmental factors [80, 81, 85, 86, 87, 96, 345–350].

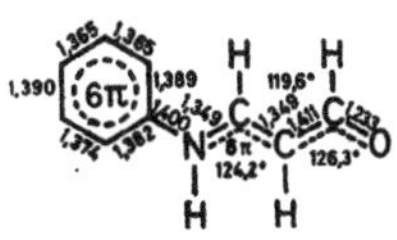

(4.1) [351]

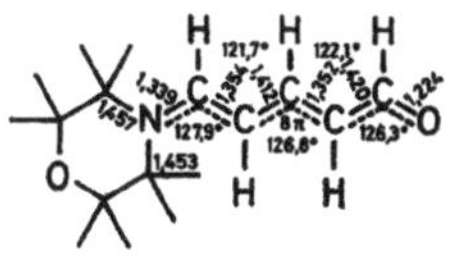

(4.2) [352]

The simple merocyanines (4.1) and (4.2) containing a tri- or a pentamethine structural group were subjected to a detailed investigation by Kulpe et al. [351, 352]. The structure analyses yielded a strong alternation of the C—C bond lengths along with very short C=O and C—N bonds. For the structure-theoretical interpretation of this finding with the aid of resonance theory we must take into account in addition to the limiting structures (4.3a) and (4.3b), as they are commonly used, polar limiting structures of the kind exemplified in (4.3c) and (4.3d) as already mentioned in Chapter 2. Otherwise one might arrive at wrong conclusions as regards the structural features of this class of compounds [353].

$$R_2\overline{N}-(CH=CH)_n-CH=\underline{O}, \qquad R_2\overset{\oplus}{N}(=CH-CH)_n=CH-\overline{\underline{O}}^{\ominus}$$

(4.3a) (4.3b)

$$R_2\overline{N}-(\overset{\oplus}{C}H-\overset{\ominus}{\overline{C}}H)_n-CH=\overline{\underline{O}}, \qquad R_2\overset{\oplus}{N}=(CH-\overset{\ominus}{\overline{C}}H)_n-CH=O$$

(4.3c) (4.3d)

Theoretical calculations based on *MO* theory [348, 354, 355] have led to different results; the calculations using the *PPP* [354] or *VESCF* formalism [348] agree best with the experimental bond lengths. That we are concerned here, in spite of the polyenic bond alternation, with substances exhibiting certain polymethine-like features is proved by the bond angles along the polymethine chain in (4.1) and (4.2) [353, 356] which, as in the polymethinecyanines (cf. Table 2), alternate distinctly. Compared to the cyanines, the angles show, on the whole, a slight shift towards higher values. The merocyanines intermediate position in terms of structural and spectroscopic properties between the ideal polymethines and the ideal polyenes is demonstrated by means of a microstructural model of solvatochromism based on the Molecular-Orbital (*MO*) theory [348, 357].

4.2. Monomethine Structural Groups

4.2.1. Merocyanines

The acid amides in (4.4) are monomethine merocyanines, for which more recent *X*-ray data have been compiled in Table 7.

Table 7: Bond Lengths [Å] and Angles In Acid Amides (4.4)

Substance	R^1, R^2	R^3	Ref.	C—O	C—N	∢ O—C—N	C—R^3
Acetamide	H, H	CH_3	[358]	1.260	1.334	123.1°	1.505
Monofluoracetamide	H, H	CH_2F	[359]	1.254	1.319	124.0°	1.533
Ammoniumoxamide	H, H	CO_2NH_4	[263]	1.240	1.324	124.2°	1.564
Malonamide	H, H	$-CH_2-$	[360]	1.254	1.334	122.6°	1.507
Glutaramide	H, H	$-(CH_2)_3-$	[361]	1.22	1.34	122°	1.52
Adipamide	H, H	$-(CH_2)_4-$	[362]	1.23	1.33	123°	1.49
Suberamide	H, H	$-(CH_2)_6-$	[363]	1.25	1.32	123°	1.52
Acelainamide	H, H	$-(CH_2)_7-$	[364]	1.234	1.325	122°	1.513
N-Methyldipropyl-acetamide	H, CH_3	C_7H_{15}	[365]	1.239	1.320	121.2°	1.510
Averages				1.242	1.327	122.8°	1.518
Averages for polypeptide chains [366]				1.23	1.32	122°	1.53

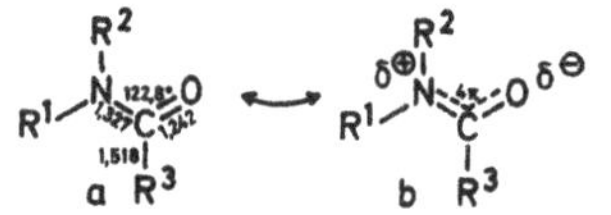

Averages taken from Table 7 (4.4)

The averages of the bond lengths of the amides coincide approximately with those determined for the acid amide grouping in polypeptides [366]. The C⎯N bond length of N-methylacetamide (4.5) deviates significantly from the average value. It may be that the polar structure b of (4.4) becomes more stabilised, owing to the inductive effect of the methyl group.

The analogous N-methyldipropyl acetamide (Table 7) or the cyclic N,N′-dimethyl-diketopiperazine (4.6) or the hexahydropyrimidine-4,6-dione [369], however, display normal acid amide distances, so that one has to look out for more structural evidence. On the other hand, there is a distinct lengthening of the C⎯N bond with simultaneous shortening of the C=O bond when the amide group is part of two competing methine fragments, as in the succinimide (4.7), the pyrrolidine-2,5-diones [371, 372] and tetra-

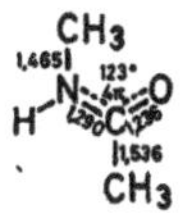

(4.5) [367]

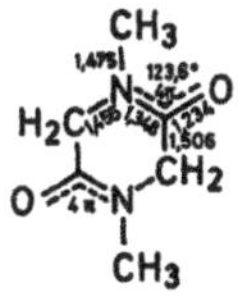

(4.6) [368]

hydrophthalimides [373, 374]. The effects are further intensified when the π-electron system is in addition caused to branch over the nitrogen atom. In the case of the N,N'-bisuccinimidyl (4.8) the C=O bond length is less than that of normal carbonyl groups (Table 1). The C=N bond of the amide groups is extended to 1.39 Å and the linking N—N bond, at 1.37 Å, is shorter than normal N—N single bond lengths (Table 1), which indicates again branching of the conjugation.

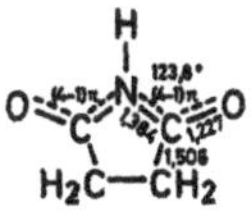

(4.7) [370]

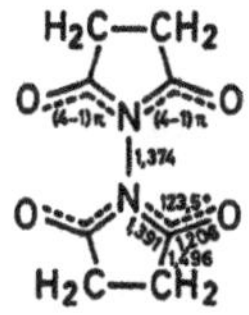

(4.8) [375]

N-phenyl substitution of the amide group in (4.9) or (4.10) likewise causes the C=N bond to be enlarged and the C=O bond to be reduced as a result of the participation of the phenyl group in the conjugation.

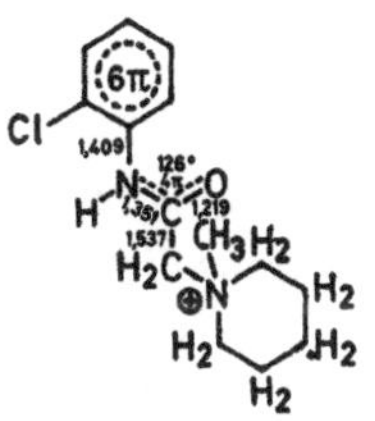

(4.9) [376]

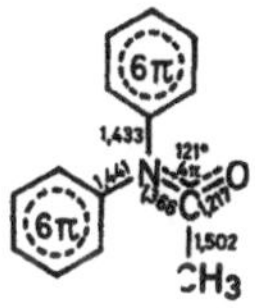

(4.10) [377]

The fact that the N-phenyl groups are involved in the monomethine resonance is indicated by the N-phenyl distances being clearly shorter than in the single bond (Table 1); here, we may probably distinguish between the effect of a single phenyl substitution in (4.9) and a two-fold substitution in (4.10).

In urea (4.11) [70, 378] and its derivative (4.12) the methine structure itself is branched, with two amide groups competing for one carbonyl group. This causes the C=O bond to be enlarged, whilst the C=N bond is only slightly extended. Influencing the conjugated system by N-phenyl substitution in (4.12) or by N-amino substitution in the

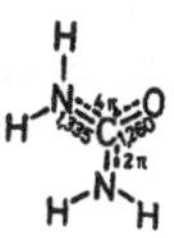

(4.11) [70]

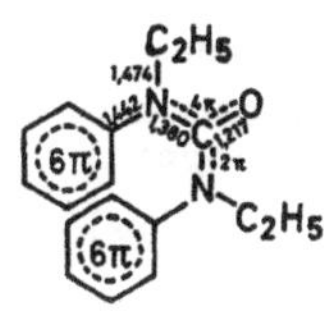

(4.12) [379]

carbamylhydrazine complexes (4.13) leads to the same alterations in the electron distribution as in the normal acid amides (4.9) and (4.10).

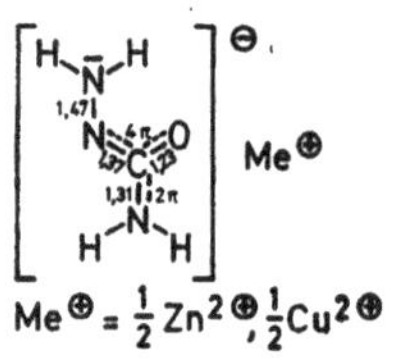

(4.13) [380]

The cyclotrimethylene-trinitramine [381], the 3,7-dinitro-1,3,5,7-tetraazacyclooctane [382] and N,N'-dinitro-ethylenediamine (4.14) may be regarded as aza-substituted monomethine merocyanines. In (4.14) the bonds involved in conjugation show the expected equalisation of the spacings of the single and double bonds (cf. Table 1), while the N—O bond lengths correspond well with those of the symmetrical azaoxonols (3.32) and (3.33). Bond length equalisation is less pronounced in the cyclotrimethylene-

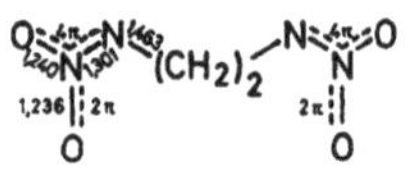

(4.14) [383]

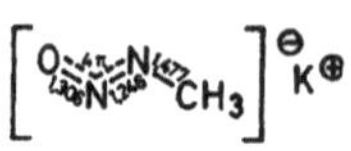

(4.15) [384]

trinitramine, with an average 1.21 Å for the N—O bond and 1.38 Å for the N—N bond [381]. Of interest is the bonding in potassium syn-methyldiazotate (4.15) which, by the polymethine concept, is to be classified as an anionic aza-monomethinemerocyanine. Here the delocalisation of the π-electron system effects a reversal even of the bond lengths, between N—N and N—O.

4.2.2. Thia-merocyanines

A typical thiamerocyanine structural group is realised in the thioacetamide (4.16). The polymethine resonance is amplified by the formation of anions. In trans-dichloro-tetrakis(thioacetamide)-nickel (II) the bond lengths are reduced, on the average, to 1.307 Å (N⩦C) and 1.680 Å (C⩦S), respectively [386]. Substitution of a hydrazide group in bis(N,N-dimethyl)thioformohydrazidato-copper(II) (4.17) for the amide group in the thioacetamide causes the C⩦N distance of 1.32 to be shortened to 1.29 Å, apparently due to an N—N interaction, as indicated by the relatively short N—N distance of 1.41 Å compared to the 1.46 Å of the N—N single bond.

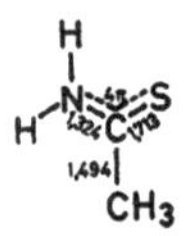

(4.16) [385]

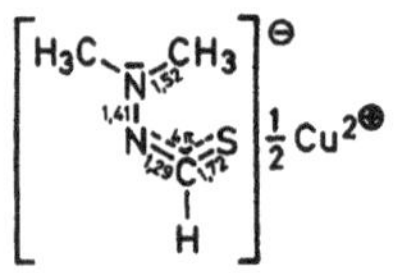

(4.17) [387]

Branched thiamerocyanine structural groups of almost identical bond lengths include the dithiocarbamates (3.36) mentioned in Section 3.3. Branching of the type observed in urea occurs in dithiobiurea [388] and in its metal complexes [389—391] as well as in the metal complexes of thiourea itself (4.18), in which an even more polymethine-like structure, especially shortening of the C=N bond lengths in the monomethinecyanine fragment is effected by means of complex formation. The data of a few more recent structures of thiourea and its complexes are summarised in Table 8. Also to be mentioned

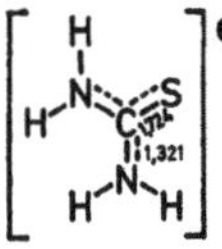

Average bond lengths taken from Table 8 (4.18)

in this context are the bis(thiourea)iodine(I) iodide [403] and the dication of α,α'-dithiobisformamidine [125] already cited in Section 3.1 under Monomethinecyanines. The relatively short C—S distances of about 1,72 Å indicates a branching of the symmetrical cyanine structure in these compounds as shown in formula (4.18). The last mentioned thiosemicarbazides in Table 8 may be regarded — similar to the analogous (4.13) — as N-amino-substituted thiourea, in which an even stronger delocalisation of the π-electron system is indicated by the N—N and C=N—(N) bonds being shorter than those in (4.13). Other compounds of a comparable electronic structure are to be found in [404] and [405].

Table 8: Bond Lengths [Å] In Thiourea, Its Metal Complexes (4.18) and In Thiosemicarbazide Derivatives

No.	Substance	*Me*	Ref.	C=S	C=N		
1	Thiourea	—	[392]	1.720	1.340	a) Average bond lengths from the literature	
2	Thiourea[a])	—	[393]	1.738	1.342		
1	Dichlorobis (tetramethyl-thiourea)copper(II)	Cu	[394]	1.735	1.33		
2	Bis(thiourea) silver(I)chloride	Ag	[395]	1.709	1.321		
3	Trans-dichloro-tetrakis(thiourea)cobalt(II)	Co	[396]	1.73	1.30		
4	Dichlorotetrakis-(trimethylenethiourea)nickel(II)	Ni	[397]	1.705	1.33		
5	Trans-dithiocyanato-tetrakis-(N,N'-diethylthiourea)nickel(II)	Ni	[398]	1.733	1.323		
6	Tetrakis(thiourea) palladium(II)chloride	Pd	[399]	1.73	1.32		
	Averages			1.724	1.321		
						C=N(—N)	N—N
1	Thiosemicarbazide	—	[400]	1.707	1.316	1.326	1.411
2	Thiosemicarbazide	—	[401]	1.713	1.331	1.334	1.415
3	Tris(thiosemicarbazide) nickel(II)dichloride monohydrate	Ni	[402]	1.698	1.329	1.330	1.419

There is evidence of an interaction between three monomethine fragments with different terminal groups in 3-methyl-4-oxo-1,3-thiazine-2-thione (4.19), where the lengths of the two C=S and the C=N bonds lie between those of a single and a double bond.

Finally to be mentioned are the isothiocyanates (4.20), whose meropolymethine-like structure is comparable to the cyano compounds (3.26) and (3.27).

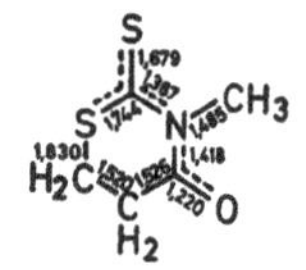

(4.19) [406]

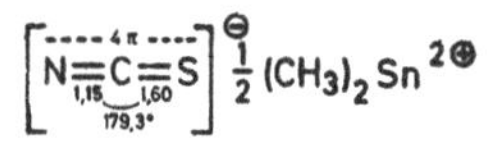

(4.20) [407]

Owing to the π-electron delocalisation, the C—S bond shows the character of a double bond without, however, enlarging the C≡N triple bond as noted earlier in the cyano-polymethines (Section 3.1).

4.2.3. Unsymmetrical Monomethineoxonols

Protonation of the monomethineoxonols (carboxylate ions) in the carboxylic acids (4.21) gives rise to an unsymmetrical π-electron distribution, but does not yield bond lengths equivalent to the 1.37 Å of the $C_{sp}2$—O single bond, so that these compounds must be ascribed a certain polymethine character according to formula (4.21). Especially

Averages taken from [408] (4.21)

the well-known enlargement of the C—C—O angle from 120° to about 126° [68] supports the assumption of a strong charge deficiency on the methine-carbon atom. Noteworthy is the short C=O distance of 1.21 Å which is analogous to the relatively short C=O bond lengths in the longer-chain simple and coupled merocyanines (Sections 4.3 and 6.2).

4.3. Trimethinemeropolymethines

Intermediate states between a highly symmetrical trimethine structure and a more or less pronounced polyene structure are realised in derivatives of tetronic acid [409, 410], ascorbic acid [411], in the enols of the malonic dialdehyde [412—415] as well as in some chelates of acetylacetone. Consequently, the unsymmetrical trimethineoxonols may be classified, analogously to the carboxylic acids discussed previously, as meropolymethines. The polymethine chain in the first-mentioned compounds, including the unsubstituted or the 2-position substituted malonic dialdehyde, has an all-trans structure. In the sterically unhindered malonic dialdehyde derivatives the bond angles alternate between 116° and 126°, this being evidence of a polymethinic charge alternation. The bond lengths of α-methyl tetronic acid are given in (4.22) as an example. Compared to the standard values (Table 1) both the C_1-C_2 distance as well as the $C_3-(OH)$ distance are significantly shorter, confirming a conjugation as indicated in the formula. Here, conjugation partly affects the C=O double bond. An extension up to 1.28 Å is observed in the malonic dialdehydes as indicated in Table 9, Nos. 1 to 4.

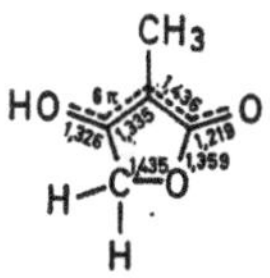

(4.22) [409]

Anderson [409] found C—(OH) bond length and acidity (acid dissociation constant) in these and similar substances to be inversely proportional. Obviously, Pauling's fundamental principle (cf. Section 2.) that the sum of bond orders on an atom remains constant has application here as well. The more strongly the oxygen is incorporated in the trimethineoxonol resonance, the more loosely bound is the proton.

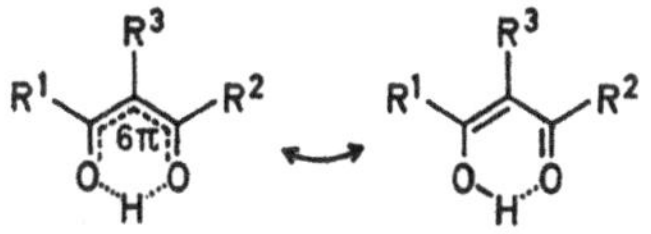

(4.23)

The enol of the 1,3-diphenyl substituted malonic dialdehyde and the acetylacetone (4.23) like their symmetrical counterparts (cf. Section 3.2) have a di-cis structure.

As in symmetrical cyanines, the anions of acetylacetone may also suffer some loss in the symmetry of electron distribution both by an unsymmetrical substitution of the methine chain and by intermolecular interactions within the crystal lattice as shown by the examples in Table 9, Nos. 5 to 12. Here the resonance within the trimethine chain is still of such a degree that bond length equalisation continues into the carbonyl group which is clearly enlarged in consequence.

A more pronounced asymmetry in the electronic structure of the acetylacetonates is obtained when substituting an imino group for the carbonyl group. The results are merocyanine-like trimethines (4.24) as compiled in Table 10.

Here, too, belongs the structure of a monothioacetylacetonatopalladium complex [428] wich was solved recently.

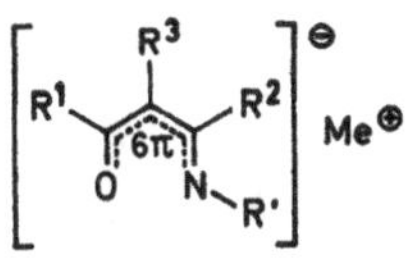

(4.24)

In the acetylacetone iminate complexes the bond length alternation is reversed in such a manner that the C=N bond exhibits a higher double-bond character than the C=O bond. Thus, an unexpectedly high double-bond character is induced in the methine chain in the carbon-carbon bond adjacent to the C=O bond. The original enamino ketone structure is retained only in the metal-free compounds enumerated in Table 10 under Nos. 5 and 6. But here the carbonyl group is not a necessary condition for the formation of this type of structure, as is apparent from a comparison with propenedianile

Table 9: Bond Lengths [Å] In Unsymmetrical Acetyl Acetonates (4.23) and Parent Compounds

No.	Substance	Ref.		R^1	R^2	R^3	*Me*
1	Énol of 2-brom-malonic dialdehyde	[412]	$O \overset{}{\underset{1.28}{=}} \underset{R_1}{C} \underset{1.38}{-} \underset{R_2}{C} \underset{1.38}{=} \underset{R_3}{C} \underset{1.33}{-} OH$	H	Br	H	—
2	Enol of 2-hydroxy- malonic dialdehyde	[413]	$O \underset{1.244}{=} \underset{R_1}{C} \underset{1.423}{-} \underset{R_2}{C} \underset{1.356}{=} \underset{R_3}{C} \underset{1.327}{-} OH$	H	OH	H	—
3	Enol of 1,3-diphenyl-1,3-propanedione	[414]	$O \underset{1.287}{=} \underset{R_1}{C} \underset{1.408}{-} \underset{R_2}{C} \underset{1.382}{=} \underset{R_3}{C} \underset{1.304}{-} OH$	C_6H_5	H	C_6H_5	—
4	do.	[415]	$O \underset{1.292}{=} \underset{R_1}{C} \underset{1.423}{-} \underset{R_2}{C} \underset{1.385}{=} \underset{R_3}{C} \underset{1.317}{-} OH$	C_6H_5	H	C_6H_5	—
5	Bis-acetylacetonato-(4-aminopyridine)-copper(II)	[275]	$O \underset{1.248}{=} \underset{R_1}{C} \underset{1.413}{-} \underset{R_2}{C} \underset{1.390}{=} \underset{R_3}{C} \underset{1.268}{-} O^{\ominus}$	CH_3	H	CH_3	Cu
6	Potassium-bis-(acetylacetonato)chloro-platinate(II)	[416]	$O \underset{1.26}{=} \underset{R_1}{C} \underset{1.44}{-} \underset{R_2}{C} \underset{1.38}{=} \underset{R_3}{C} \underset{1.35}{-} O^{\ominus}$	CH_3	H	CH_3	Pt
7	Bis[3-phenyl-acetylacetonato]copper(II)	[417]	$O \underset{1.248}{=} \underset{R_1}{C} \underset{1.454}{-} \underset{R_2}{C} \underset{1.438}{=} \underset{R_3}{C} \underset{1.263}{-} O^{\ominus}$	CH_3	C_6H_5	CH_3	Cu
8	Bis[1-phenyl-1,3-butanedionato]oxo-vanadium(IV)	[418]	$O \underset{1.295}{=} \underset{R_1}{C} \underset{1.421}{-} \underset{R_2}{C} \underset{1.388}{=} \underset{R_3}{C} \underset{1.292}{-} O^{\ominus}$	CH_3	H	C_6H_5	VO
9	Bis[1-phenyl-1,3-butanedionato]copper(II)	[419]	$O \underset{1.244}{=} \underset{R_1}{C} \underset{1.406}{-} \underset{R_2}{C} \underset{1.397}{=} \underset{R_3}{C} \underset{1.282}{-} O^{\ominus}$	CH_3	H	C_6H_5	Cu
10	Bis[1-phenyl-1,3-butanedionato]zinc(II)-monoethanolate	[420]	$O \underset{1.286}{=} \underset{R_1}{C} \underset{1.412}{-} \underset{R_2}{C} \underset{1.382}{=} \underset{R_3}{C} \underset{1.278}{-} O^{\ominus}$	CH_3	H	C_6H_5	Zn
11	Dioxo-bis-(1,3-diphenyl-propanedionato)-molybdenum(VI)	[421]	$O \underset{1.27}{=} \underset{R_1}{C} \underset{1.40}{-} \underset{R_2}{C} \underset{1.365}{=} \underset{R_3}{C} \underset{1.30}{-} O^{\ominus}$	C_6H_5	H	C_5H_5	Mo
12	Bis[1-ethoxy-1,3-butanedionato]-copper(II)	[422]	$O \underset{1.28}{=} \underset{R_1}{C} \underset{1.39}{-} \underset{R_2}{C} \underset{1.31}{=} \underset{R_3}{C} \underset{1.32}{-} O^{\ominus}$	OC_2H_5	H	CH_3	Cu

Table 10: Bond Legnths [Å] In Acetylacetone Iminate Complexes (4.24) and In Similar Trimethinemerocyanines

No.	Substance	Ref.		R^1	R^2	R^3	R'	*Me*
1	Tris[acetylacetonato]-(4-amino-3-penten-2-one) ytterbium	[288]	$^{\ominus}O \overset{1.26}{-} \underset{R_1}{C} \overset{1.38}{=} \underset{R_2}{C} \overset{1.41}{-} \underset{R_3}{C} \overset{1.35}{=} \overset{\oplus}{N}(R')_2$	CH_3	H	CH_3	H	Y*)
2	N,N'-Ethylene-bis-[acetylacetone-iminato]-oxovanadium(IV)	[423]	$^{\ominus}O \underset{1.300}{-} \underset{R_1}{C} \underset{1.366}{=} \underset{R_2}{C} \underset{1.390}{-} \underset{R_3}{C} \underset{1.295}{=} N{-}R'$	CH_3	H	CH_3	$-CH_2-$	VO
3	N,N'-Ethylene-bis-[acetylacetone-iminato]-copper(II)-hemihydrate	[424]	$^{\ominus}O \underset{1.325}{-} \underset{R_1}{C} \underset{1.365}{=} \underset{R_2}{C} \underset{1.385}{-} \underset{R_3}{C} \underset{1.335}{=} N{-}R'$	CH_3	H	CH_3	$-CH_2-$	Cu
4	Potassium N,N'-Ethylene-bis(acetyl-acetone-iminato)trans-diglycinato-cobaltate(III)-hexahydrate	[425]	$^{\ominus}O \underset{1.300}{-} \underset{R_1}{C} \underset{1,367}{=} \underset{R_2}{C} \underset{1.416}{-} \underset{R_3}{C} \underset{1.305}{=} N{-}R'$	CH_3	H	CH_3	$-CH_2-$	Co
5	2-(4'-Amino-5'-aminopyrimidyl)-2-penten-4-one	[426]	$O \underset{1.26}{=} \underset{R_1}{C} \underset{1.41}{-} \underset{R_2}{C} \underset{1.37}{=} \underset{R_3}{C} \underset{1.37}{-} NH{-}R'$	CH_3	H	CH_3	$C_4N_3H_4$	—
6	β-Anilinopropenal	[351]	$O \overset{1.233}{=} \underset{R_1}{C} \overset{1.411}{-} \underset{R_2}{C} \overset{1.349}{=} \underset{R_3}{C} \overset{1.349}{-} NHR'$	H	H	H	C_6H_5	—
7	Propenedianil	[427]	$R'N \underset{1.282}{=} \underset{R_1}{C} \overset{1.414}{-} \underset{R_2}{C} \overset{1.361}{=} \underset{R_3}{C} \overset{1.352}{-} NHR'$	H	H	H	C_6H_5	—
8	Bis[1,3-diphenyl-propane-monothiodionato]-palladium	[428]	$O \underset{1.22}{=} \underset{R_1}{C} \underset{1.42}{-} \underset{R_2}{C} \underset{1.345}{=} \underset{R_3}{C} \underset{1.745}{-} S^{\ominus}$	C_6H_5	H	C_6H_5	—	Pd

*) The asymmetric ligand of the same compound included in Table 4. In this case the ytterbium is bound to the acetylacetone ligand.

(4.25), which is terminated by two nitrogen atoms, and in which asymmetry is induced only by protonation of one terminal nitrogen atom (cf. Table 10, No. 7).

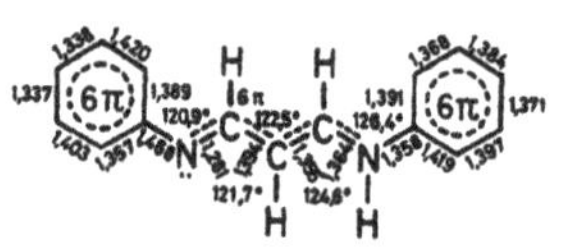

(4.25) [427]

While the 2-amino-4-thiazolinones (4.26) [214—217], the 2-(nitromethylene)piperidine [429] and the rhodan hydrate [430] exhibit typical bond lengths equalisation within the aza-substituted trimethinemerocyanine fragment, more polyene-like structures are formed, for example, in the 4-iminopentane-2,3-dione 3-oximato nickel(II) complex [431] or in N-methyl-pyridine-2-aldoxime (4.27). In the latter obviously the aromatic π-electron sextet is dominating in the structure. More substituted heterocyclic compounds are compiled in Section 5.2.

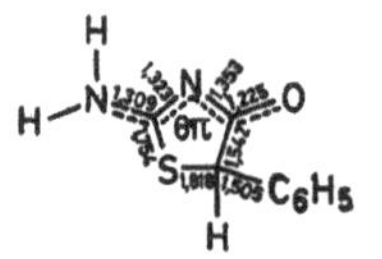

(4.26) [217]

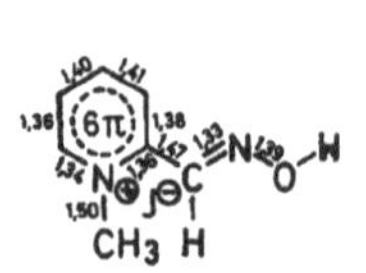

(4.27) [432]

The formazanes (4.28) must be classified as diazamerotrimethines, which, disregarding the short N=N bond on the left side of the molecule, show a relatively high degree of bond length equalisation, obviously because of similar terminal groups. The same structure but substituted in the meso position by a mercapto group is realised in S-methyldithizone [434].

A branched trimethinemerocyanine framework is contained in the 2-amino-1,1,3-tricyanopropene (4.29) whose carbon-carbon bonds are also partly equalised indicating a polymethine-like π-electron distribution, whereas an amino-substituted acrylonitrile [436] exhibits a more polyenic character by comparison.

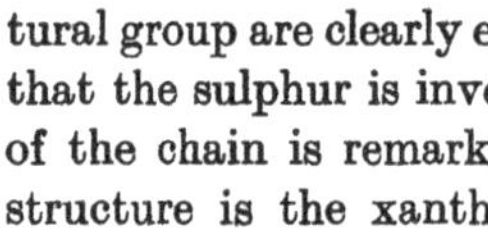

(4.28) [433]

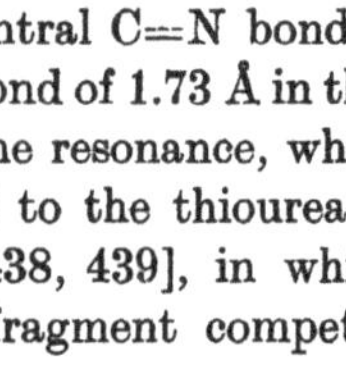

(4.29) [435]

In the thiourea derivative (4.30), an azathiatrimethinemerocyanine competes with a thio-monomethine structural group. The central C=N bonds of the trimethine structural group are clearly equalised. The C—S bond of 1.73 Å in the thiazoline ring indicates that the sulphur is involved in the trimethine resonance, whilst the C=S double bond of the chain is remarkably short compared to the thioureas (cf. Table 8). A related structure is the xanthan hydride (4.31) [438, 439], in which, as in the dithiolium salts (3.37) and (3.38), an azatrimethine fragment competes with an aromatic dithiolium ring system.

More complicated, branched trimethine and monomethine fragments are contained in the typical merocyanine dyes such as (4.32) and similar compounds [440b, 440c, 440d]. The polymethine resonance causes the $C_5—S_1$ and $C_4—C_5$ bond lengths in (4.32) to be shortened by comparison to the rhodanine ring (4.33). Probably additional interaction effects between the trimethine structural group and the S=C—N monomethine structural group in the rhodanine ring as well as a possible trend of the ring towards

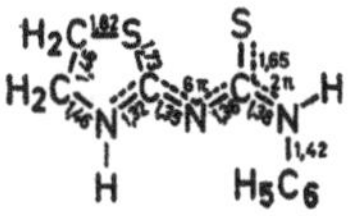
(4.30) [437]

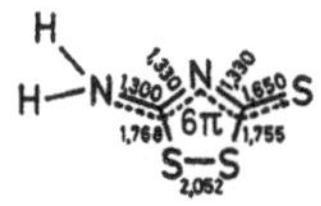
(4.31) [438]

aromatisation play a certain role in the dye (4.32). As expected, its becomes more polyene-like, thus approximating to the unconjugated reference substance (4.33), when the nitrogen atom of the thiazolin ring is replaced by sulphur [440f].

Finally, mention should be made of the structure of N-ethyl-2,2′-dimethylsulphonylvinylidene amine (4.34) and similar compounds [442], in which a trimethine resonance

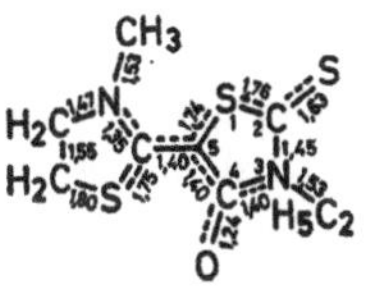
(4.32) [440a]

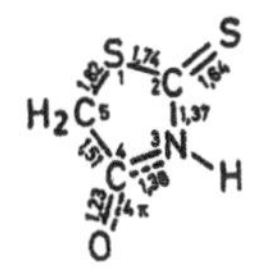
(4.33) [440e]

involving a sulphur bridge is realised (the reference values of the unconjugated bond lengths, cf. Table 1, are given in brackets). There is a shortening of the C=N double bonds towards a triple bond, to which preference seems to be given as has been pointed out in Sections 3.1 and 4.2.2.

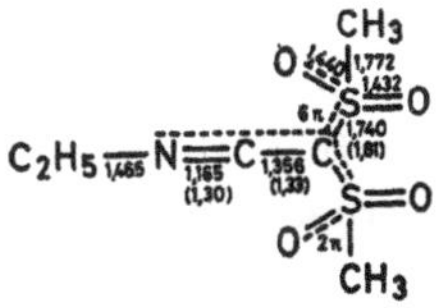
(4.34) [441]

The discussion of the aforementioned structures from the point of view of more or less pronounced intermediates between ideal polymethines and ideal polyenes in terms of formula (4.3) reveals again the limitations of the symbolism used herein. Some of the double or triple bonds are localised to a larger extent and presented as such in the formula. In relation to many compounds it remains however a matter of interpretation whether one should prefer the presentation with a delocalised π-electron system or with fixed double and single bonds. Whichever the case, one should always bear in mind the charge alternation directly associated with a more or less pronounced polymethine-like π-electron delocalisation.

4.4. Pentamethinemeropolymethines and Longer Chains

The only structure of a typical pentamethinemerocyanine chain (4.2) [352] that has been solved thus far was discussed in the introduction on polymethine-polyene intermediates (Section 4.1). Other meropolyene-like structural groups are contained in many complex molecules, in which aromatisation tendencies frequently make it impossible to unambiguously establish their structures. Such an attempt is most likely to succeed in the coupled polymethines discussed in Chapter 6. For this reason, the following structures are not in every case representative but reveal the usefulness of the polymethine concept.

In (4.35), a simple pentamethinemerocyanine competes with a trimethinemerocyanine chain through a joint nitrogen atom.

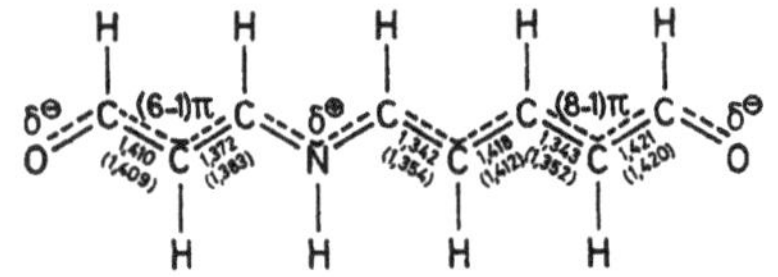

(4.35)

values above: bond lengths of (4.35) estimated from H,H-coupling constants ΔJ [443]
values below (in brackets): bond lengths of isolated structures taken from (4.2) and (6.13)

As in the respective monomethines (Section 4.2.1) the polymethine character of the competing chains should become weaker as a result. The only way to demonstrate this is by ^{1}H-nuclear magnetic resonance spectroscopic studies [443] because of the lack of suitable X-ray data. The electron density alternation derived from the chemical shifts of the protons is less pronounced in (4.35) while the bond length alternation estimated from the vicinal proton coupling constants J is greater than the respective values in the isolated structures which can be taken from (4.2) and (6.13), respectively. In estimating the alternations in bond length we used the empirically derived correlation $J[\text{Hz}] = 127.86 - 81.85\,d\,[\text{Å}]$ and $\Delta J[\text{Hz}] = -81.85\,\Delta d\,[\text{Å}]$ for all-trans configurated polyenes and polymethines given by SCHEIBE et al. [444], taking into account the inductive effect of the heteroatoms adjacent to the C-atoms on the coupling constants by using the correction term determined in [349]. It is shown that the effect on the pentamethine chain is stronger on account of its greater, experimentally proved, polarisability [86] than on the trimethine chain of the molecule.

A repeatedly branched pentamethine structural group can be formulated in the potassium 4,4-dinitro-2-butenamide (4.36). The single bond length between

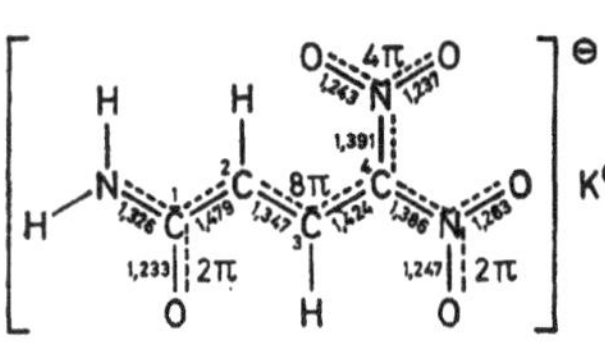

(4.36) [445]

$C_1 - C_2$ indicates however that the monomethine resonance is dominating in the amide group while there is preferentially a trimethine resonance within the dinitromethane structural group.

Accordingly, the N=O bonds are enlarged as in the complexes of Meisenheim (3.32) and other azapolymethines of Section 3.2. The fact that the double bond in the pentamethine chain between C_2 and C_3 is enlarged and the single bond between C_3 and C_4 is narrowed shows that the pentamethine resonance in terms of the structure indicated in (4.36) must not be neglected.

We find branched aza-pentamethine-merocyanine structural groups in various difurazanes [446–449], for example in the red 4-imino-8-acinitro-benzo[1,2-c:4,5-c′]-difurazane anion (4.37).

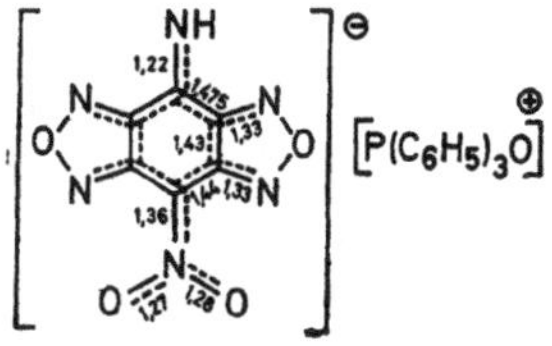

(4.37) [446]

At 1.22 Å the C=N bond of the imino group is greatly reduced in length, which is not the case in normal amino substituted furazanes [450]. This shortening is a result of branching effects in the central ring system and a strong electron shift towards the nitro groups whose N=O bond lengths have been enlarged correspondingly. The meropolymethine-like bond alternation continues in the pentamethine chain, with the three ring bonds being comparatively long because of further branching.

Formally it is possible to formulate in 1-methyl-2-picryliminoindoline (4.38) two diaza-pentamethine structures between the nitrogen atom of the indoline ring and the

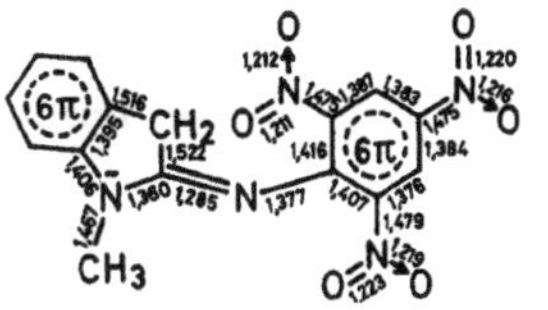

(4.38) [451]

ortho-positioned nitro groups as well as a heptamethine structure between the ring nitrogen atom and the para-positioned nitro group. Compared to (4.37) there is a more pronounced bond length alternation, indicating the preferential formation of a polyene-aromatic structure.

Based on the available structural data, a more polyene-like structure seems to be formed in the trans-β-2-furylacrylamide (4.39) as well. In principle, the formation of a heptamethinemerooxonol between the oxygen atom of the furane ring and the carbonyl-oxygen atom of the acid amide group would be conceivable. The enlarged

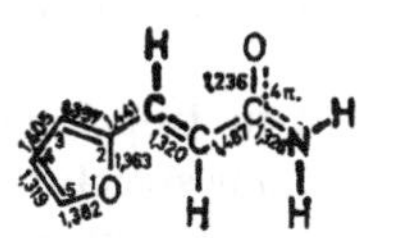

(4.39) [452]

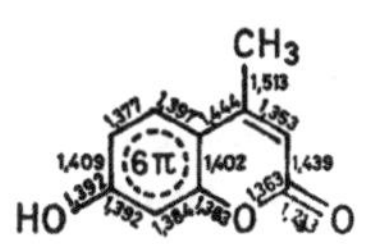

(4.40) [455]

C_2—C_3 double bond in the furane ring and the narrowed C_2—C_α single bond of the substituent are indications of this conjugation being involved to a certain extent. Principally the same considerations apply to the trans-1-(2-chloro-4-dimethylaminophenyl)-2-nitroethylene [453] and the 4-hydroxy-3-nitrocinnamate [454], both of which contain a heptamethine fragment in competition with an aromatic 6π-system. In the latter, an azatrimethinemerooxonol can be formulated in addition. One heptamethinemeropolymethine fragment each is contained in both the 4-methylumbelliferone (4.40) and the 7-diethylamino-4-methylcoumarin [456], which play an important role as active media in dye laser technology. Also the structure of a more complex coumarin dye is known [457]. With the exception of the polyenic bonds in the neighbourhood of the carbonyl group, the aromatic bonding state is seemingly preferred in the rings. One would, however, expect a stronger polymethinic character in the anions or cations of the same dyes. The same holds for fluorescein in its colourless lactoid form [458] which even contains a nonamethineoxonol fragment.

Typical polyene-polymethine intermediates may be observed in the pentafulvene series [18, 459–466], of which some examples are given in (4.41) to (4.45). As a rule with increasing electron-donor-acceptor ability of the substituents the double bond lengths of the fulvene (4.41) are enlarged and the single bond lengths reduced. In (4.42), with

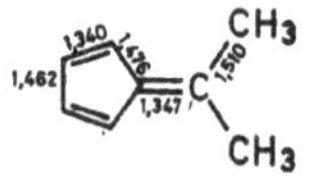
(4.41) [18]

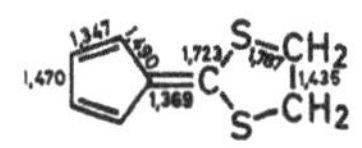
(4.42) [464]

a bond length of about 1.72 Å as compared to the 1.81 Å of a C—S single bond (Table 1), involvement of the sulphur atoms in the conjugation system is evident. The same is true of the C—N bond length in (4.43) and the corresponding bis(dimethylamino)-fulvene

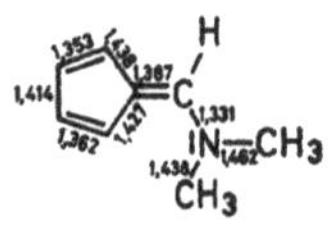
(4.43) [460]

[466b]. In (4.44) and (4.45) a mero-heptamethine structural group can be formulated.

In (4.44), the cyano group, which itself is not lengthened (cf. Section 3.1), is involved in the conjugation as indicated by the distance of 1.40 Å to the ring (the average bond lengths in acetonitrile are by comparison: $N \overset{1.156}{\equiv} C \overset{1.469}{-} CH_3$ [467, 468].

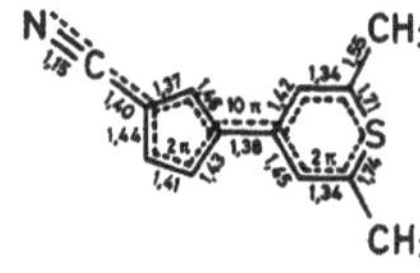
(4.44) [465]

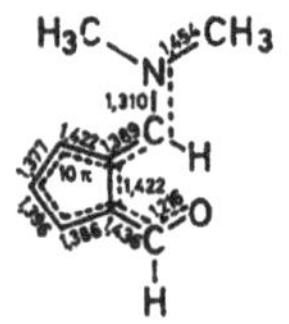
(4.45) [463]

The equalisation of bond lengths is particularly pronounced in 2-formyl-6-dimethylamino-pentafulvene [460, 463] (4.45) whose structure is hardly comparable to the polyenic fulvene any longer.

A meropolymethine-like heptamethine structural group is contained in tropolone (4.46) [469, 470]. Here, too, the transitional stages from the ideal polymethinic structure in the tropolone anion (3.34), or its cyanine analogue (3.2), to a strongly polyenic structure in 3-acidotropone (4.47), 2-chlorotropone [472], and 4,5-benzotropones [473, 474], respectively, or in the heptafulvalene [51, 475, 476] already mentioned in Chapter 1., are performed. Compared to the anion (3.34), the C—O bonds in (4.46) are shortened and lengthened, respectively, without attaining the strong bond length alter-

(4.46) [469]

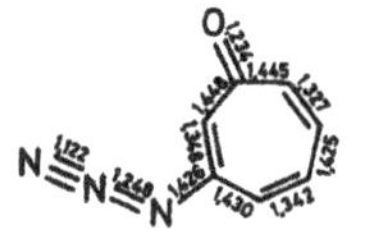
(4.47) [471]

nation observed in tropone (4.47), not to mention the values expected for a true double or single bond (cf. Table 1). The C_2—C_3, C_4—C_5 and C_6—C_7 single bonds of tropolone undergo a shortening and the double bonds C_1—C_2 and C_5—C_6 are lengthened.

Further suitable objects for testing the polyene-polymethine intermediates present are the stilbenes, azobenzenes and similar compounds, provided a polymethine-like electron

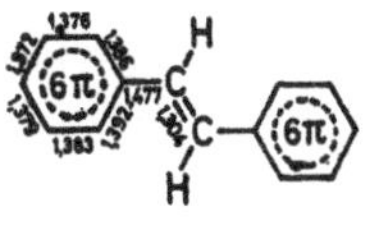

(4.48) [477]

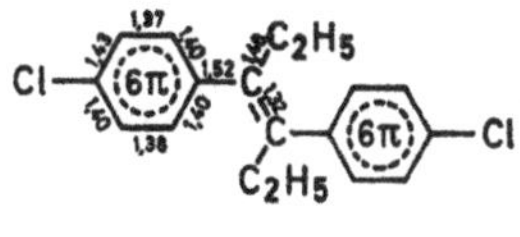

(4.49) [478]

delocalisation is induced in these substances by substitution at opposite ends with electron donor or acceptor substituents. For comparison, a few compounds with a typical polyene-like structure are presented in (4.48) to (4.51). More structures of similar compounds are to be found in [70, 481—493].

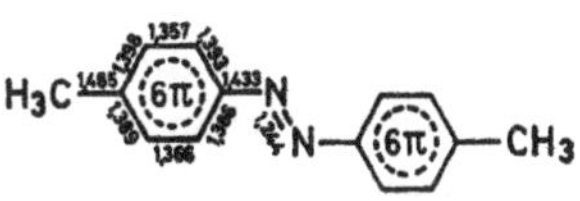

(4.50) [479]

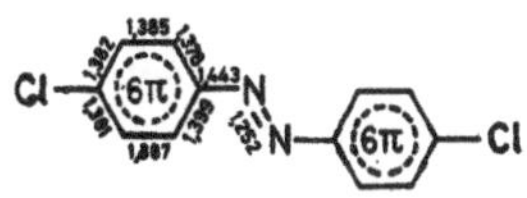

(4.51) [480]

At about 1.48 Å the C_{sp^2}—C_{sp^2} bonds (and 1.44 Å for the C_{sp^2}—N_{sp^2} bonds) within the central chain exhibit typical single-bond lengths, whereas the C=C and N=N bonds with 1.32 and ≦ 1.25 Å, respectively, represent true double bonds. This polyene-like bond alternation is indicated to continue weakly in the rings, where a bond length equalisation (alternately 1.40 and 1.38 Å) is observed, owing to the tendency to aromatisation.

As the result of unsymmetrical substitution a meropolymethine-like structure is formed in the compound (4.52) on which only inaccurate structural data are available.

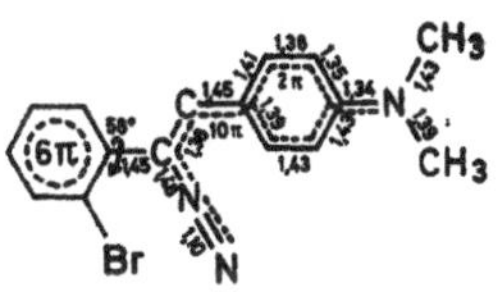

Dihedral angle 58° (4.52) [494]

Because of the twisting of the phenyl ring at the central C-atom of the chain, resonance apparently occurs preferentially between the dimethylamino group and the cyano group which is again not enlarged. In accordance with this, the central C=C double bond length is enlarged to 1.38 Å and the C—N bond length between the dimethylamino group and the ring is shortened to 1.34 Å.

In spite of a certain twist of the planes of the two phenyl rings, polyene-polymethine intermediate stages can be observed in the azomethines (4.53) to (4.55) and in [496].

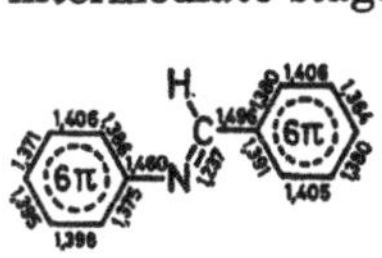

(4.53) [495]

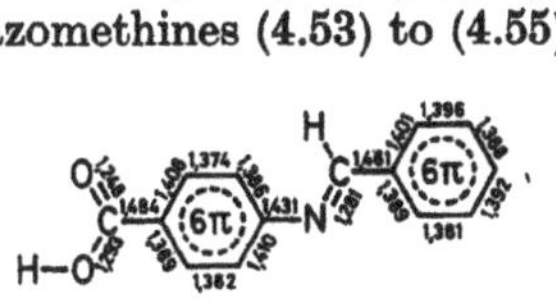

(4.54) [495]

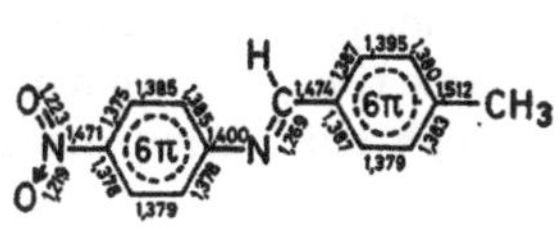

(4.55) [495]

The average C=N bond lengths increase from 1.24 to 1.28 Å, whilst the adjacent C—N and C—C distances to the phenyl rings decrease. The effect of a polymethine resonance persists also in the carboxy group whose C—C and C—O distances are shortened compared to simple carboxylic acids (cf. Section 4.2.3) while the C=O bond length is somewhat enlarged. The same tendency towards an equalisation of the C—N bond lengths can be observed in the pyrazoline azomethine dyes (4.56). The example shows the azomethine bond being enlarged to 1.30 Å as is to be expected, accompanied by a shortening of the C—N bond lengths to the phenyl ring to 1.37 and 1.36 Å, respectively.

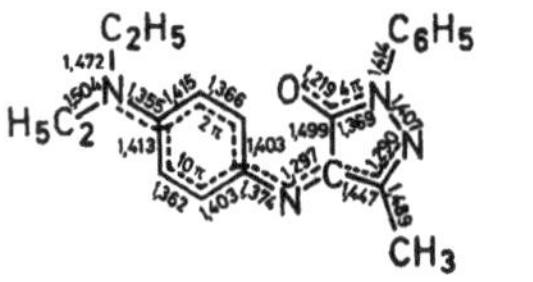

(4.56) [497]

In the azo dye methyl orange (4.57), too, the two C—N single bonds of the dimethyl-amino-substituted ring at 1.41 and 1.36 Å, respectively, are shorter than in the azobenzenes or corresponding aminobenzenes (cf. Table 1), indicating the formation of a certain polymethine-like structure shown in the formula.

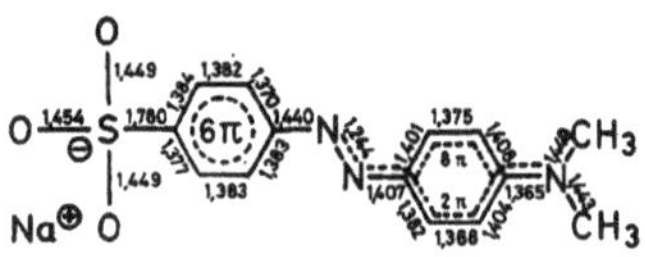

(4.57) [498]

However, an unsymmetrical substitution of the polyenes need not necessarily induce a polymethine-like structure, as indicated by the largely polyene-like compounds (4.58) and (4.59). Here the acceptor and donor strengths of the opposite substituents are

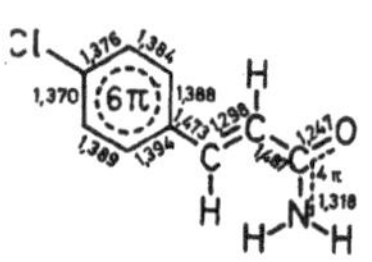

(4.58) [499]

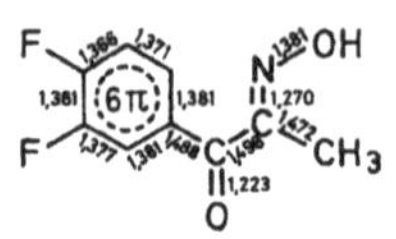

(4.59) [500]

apparently too small to cause a delocalisation of the π-electrons over the entire molecule.

A likewise more polyene-like state is to be found in the coloured compounds (4.60) to (4.62), with indications of a quinone structure in the dienone ring, particularly in (4.60). The bond length equalisation in the order (4.60) < (4.61) < (4.62)

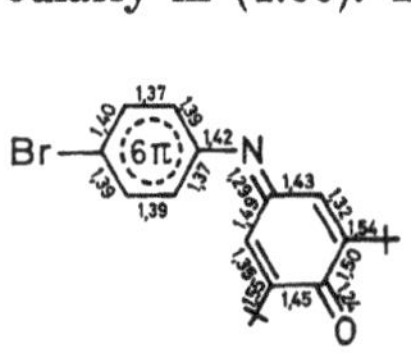

(4.60) [501]

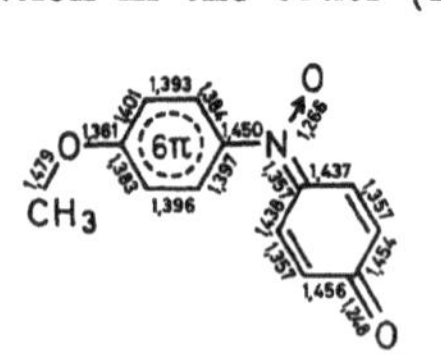

(4.61) [502]

indicates the transformation to the polymethinic structure, although the structural data in part were determined without great accuracy. At the same time, the aromatisation tendency of the rings may have a certain part in it, though not a dominating one, as is evident when comparing (4.62) to the diazonium ion (4.63), whose ring bonds are more equalised.

The structure of the diazonium ion should be modified in the direction of a more polymethine-like electron distribution by para-substitution with an amino group. This

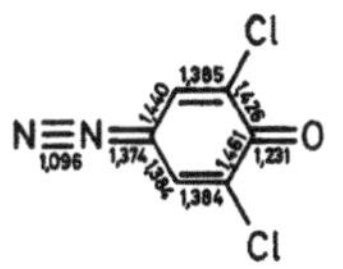

(4.62) [503]

(4.63) [504]

seems to be confirmed by the N≡N bond in the para-dimethylaminodiazonium ion [505] being enlarged from 1.10 to 1.17 Å, but the structure analysis is not accurate enough to allow a detailed discussion.

The deep-blue pyridinium betaine (4.64) is strongly negative solvatochromic, that is, in solution its longest wavelength absorption band is shifted towards shorter wavelengths as the polarity of the solvent increases.

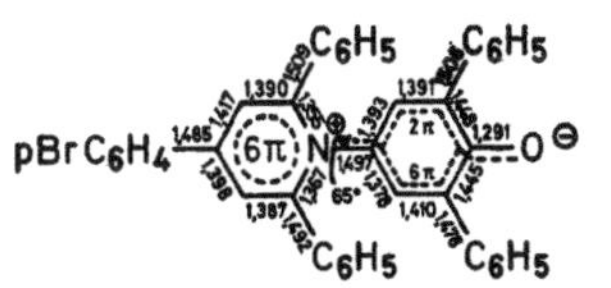

(4.64) [506]

According to the current concepts of solvatochromism [357, 507, 508], a strongly dipolar betaine structure as in formula (4.3b) should already be present in dyes of this type in non-polar solvents, and its stability increases with increasing solvent polarity. In the crystalline state the dye molecule has in fact an abnormally long C=O bond of 1.29 Å which points to a high electron density on the oxygen atom. A corresponding shortening of the C—N bond length between the two central rings is missing, however. This may be attributed to branching and twisting effects. Moreover, the fact that an even number of methine atoms is present between the two terminal heteroatoms certainly is of some importance, so that resemblances to the polymethine radicals (Chapter 7) should appear here.

5. Polymethine-Aromatic Intermediates

5.1. Substituted Benzenes

As has been already outlined, benzene and all higher condensed aromatic hydrocarbons can be gradually converted to polymethine-like, deeply coloured compounds by substitution with electron donors and acceptors. There is a large body of *X*-ray analyses of substituted aromatics, so that it suffices to point out a few typical examples.

When we consider the series benzene (5.1), aniline (5.2), para-aminobenzaldehyde (5.3), and its immonium derivative (5.4) we note to an increasing extent polymethine-

(5.1) [10]

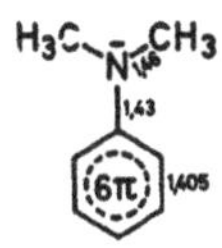

(5.2) [509]

like features, deepening of colour, facilitation of electrophilic and nucleophilic reactions, etc. [9]. Formally an aromatic π-sextet can be formulated in the compounds (5.1) to (5.4), while in (5.3) and (5.4) in addition branched polyenic and polymethinic fragments, respectively, can be formulated. In (5.2) already participation of the amino group in the

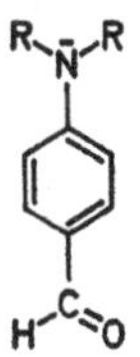

(5.3)

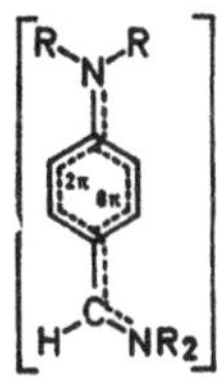

(5.4)

resonance indicates a moderate polymethinic character with a total of 8π-electrons on 7 atomic centres. The high degree of localisation of two π-electrons at the nitrogen atom, however, is responsible for the dominating aromatic character in (5.2), whilst in (5.3), in analogy to the merocyanines (Chapter 4.), an increasingly polymethinic character is to be expected. The uniformity of π-electron distribution is greatest in (5.4), without yet attaining the ideal polymethinic structure observed in the pentamethinecyanines (3.1, $n = 5$). The ideal structure is modified by the double branching of the polymethine chain and by the ring's tendency of aromatisation. The only structure available of (5.3) is that of the hydrobromide [510], in which the conjugation of the amino group to the ring is eliminated. Up to now no *X*-ray structure analysis has been carried out of (5.4). The chemical and physico-chemical behaviour of (5.4) as compared to the simple pentamethinecyanines (3.1, $n = 5$), such as higher sensitivity to hydrolysis and a somewhat lower π-electron density alternation, as evidenced by the ^{1}H-*NMR* spectra, seem however to confirm the expected structural features.

Using structures (5.5) to (5.8) a closer study is afforded of the intermediates between the aromatic state and the ideal polymethine state. As the acceptor strength of the

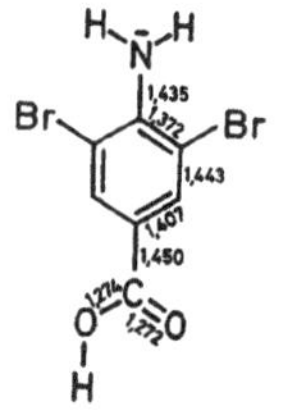

(5.5) [511]

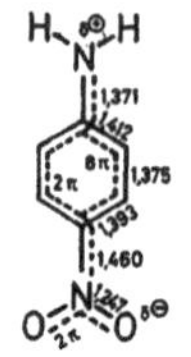

(5.6) [512]

para-positioned substituent increases, the N-phenyl single bond of the amino group becomes shorter and the ring bonds at the branching points are extended, as a consequence. Caused by the branching effect also, polyene-like structural groups are involved.

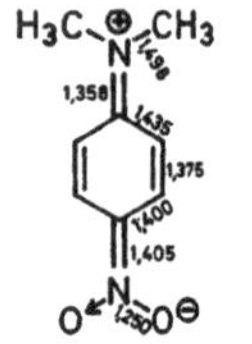

(5.7) [513]

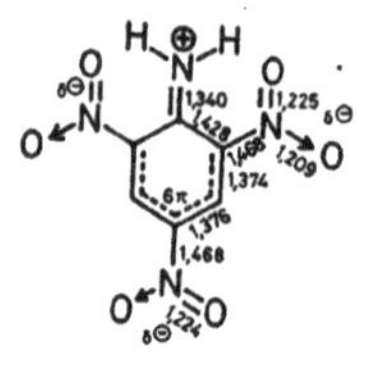

(5.8) [514]

Bond alternation originates in (5.5) from a non-polar limiting structure, and is reversed in (5.6) and (5.7), similarly to some trimethinemeropolymethines (Section 4.3, Table 10), as a result of the strong involvement of the polar resonance structure presented herein. Correspondingly, the N-phenyl bond of the nitro group in (5.6) and (5.7) is shorter compared to the nitrobenzenes [515—519] and the N—O bonds are enlarged (cf. Table 1). In (5.8), on the other hand, the electronic charge of the amino group which is placed uniformly at the disposal of the three nitro groups, further reduces the bond length of the amino group, while the N-phenyl bond lengths of the nitro groups nearly re-approach the normal value of 1.48 Å of *s*-trinitrobenzene [520]. This effect can be compensated in (5.9) and (5.10) by introducing more amino groups, in

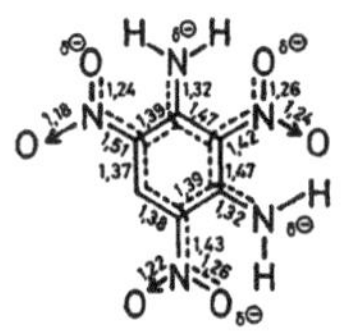

(5.9) [521]

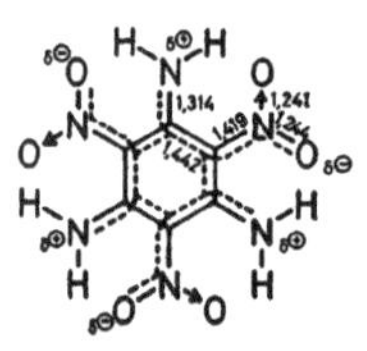

(5.10) [522]

which case trimethine structural groups may be formulated between the ortho-positioned nitro and amino groups which will amplify the polymethine resonance. This emphasises the branching effects, leading to a further enlargement of all ring bonds, especially of those which are connected with the centrally located nitro group. Also, the NH_2-phenyl bond undergoes further shortening which can be reversed, as expected, when substituting a strong electron acceptor at the amino group itself as it happens in the N-methyl-trinitroacetanilide [523], in the N-fluoroamino substituted trinitrobenzene [524, 525] or in the tetranitroanilines [526, 527]. The N-phenyl bond will then be again $\geqq$ 1.4 Å. On the other hand, the NO_2 phenyl bonds may also be enlarged to an average bond length of 1.48 Å by twisting the nitro groups out of the molecular plane by means of bulky triphenylphosphine substituents on the amino group [528].

Analogous effects can be observed in the para- and ortho-nitrophenols [529—534]. Being only slightly modified by the branching effects, the largely aromatic structure of the 2-chloro-4,6-dinitrophenol (5.11), or of the parent 2-bromo-compound [531], for example, shows distinct changes towards an azatrimethine structural fragment on anion formation like in the potassium-ortho-nitrophenolate hemihydrate (5.12), for instance.

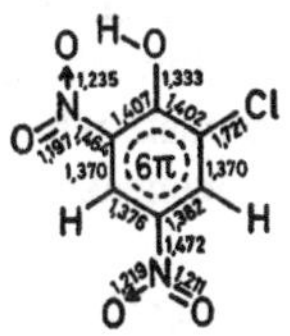

(5.11) [529]

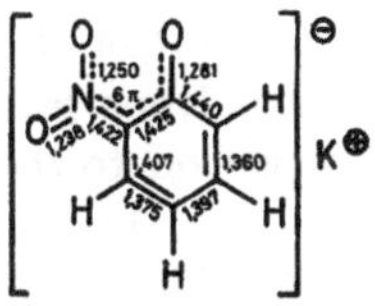

(5.12) [530]

In this context the structures of meta substituted benzene derivatives will be of interest, in which — by the classical concepts — there should be no resonance between the substituents. This seems to be confirmed for instance by the structure of m-nitrophenol (5.13) whose bond lengths agree well with nonconjugated benzenes. Yet, in the *X*-ray analyses of meta-amino-substituted nitrobenzenes [536—538] there is evidence of a slight shortening of the N-aminophenyl bond, as shown in (5.14).

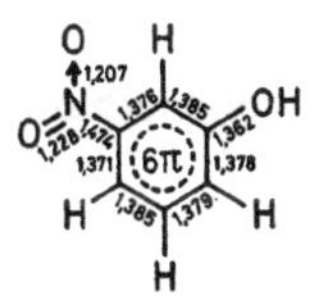

(5.13) [535]

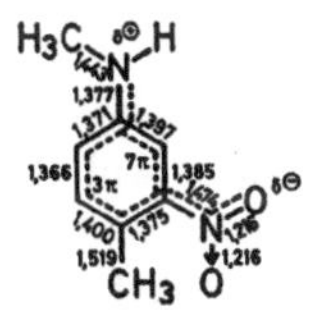

(5.14) [536]

According to the polymethine concept, these compounds allow branched polymethine structural groups with an e v e n number of methine groups (tetra- and hexamethines) to be formulated which formally are to be assigned to the polymethine radicals (Chapter 7.). These are no true radicals, however, because of branching. The shortening of the bonds favours an interpretation of the structure in this sense while the relatively high degree of asymmetry of π-electron distribution along the methine chain apparently prevents the arrangement of an ideal polymethine structure. Further investigations are necessary, however, to give a definite explanation of the structure of meta substituted aromatics.

A competition between a trimethine fragment and an aromatic 6π-ring system occurs also in the Schiff's bases of the salicylaldehyde and its metal chelates. According to formula (5.15) the interaction between the trimethinemerocyanine structural group and the orthosubstituted ring induces in the aromatic ring a polyene-like bonding which will be the more pronounced, the more developed the polymethine structure. The great number of X-ray analyses published has been evaluated earlier [289]. Table 11 contains the data of some more recent examples, which are identical with those obtained

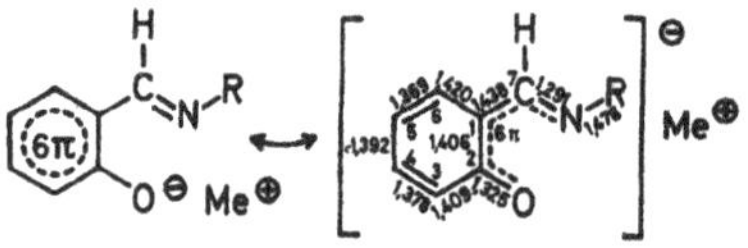

Averages taken from Table 11 (5.15)

earlier within a significance point of 3σ. More structures will be found in [551–565]. As is to be expected, the C—O bond with an average 1.32 Å is longer than the C—O bonds in the trimethine chain of the acetylacetonates (Sections 3.2 and 4.3) but obviously shorter than a C_{sp^2}—O single bond (1.37 Å). Compared to other merocyanines, the C—C bond of the aldehyde-imine group to the benzene ring is comparatively long which has to be attributed to a branching effect of the resonance system. It is certainly branching effects, too, that cause the polyene features in the benzene ring to diminish when another benzene ring is fused on to the ring, as is the case in the bis(benzeneazo-β-naphthol)-copper(II) (5.16), e.g. the aza-analogue of the salicylaldehyde imines and similar compounds [567, 568]. The relatively short C—O bond and the long N=N double bond are proof of a preferential formation of the azatrimethine structure in the chelate ring.

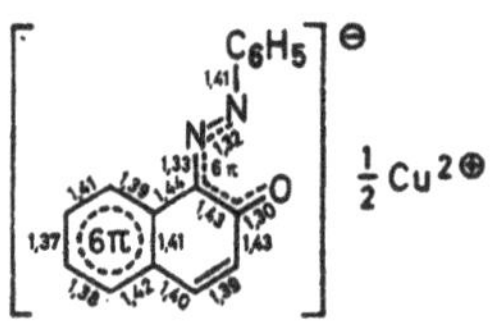

(5.16) [566]

It may be assumed that the interaction effects depicted between polymethinic, aromatic, and polyenic structures will also influence the light absorption of the salicylaldehyde imines; thus far no systematic studies have been made to this effect.

The structure of the triphenylmethylcarbenium ion (5.17) represents an example of a true carbon-polymethine exhibiting a π-electron deficit $(N - 1)$ occupation according

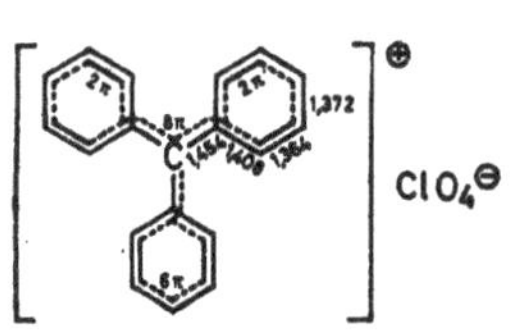

(5.17) [569]

to formula (2.1) $Y = X = X' = C$) in competition with a strong aromatisation tendency of the rings. The triple branching of the resonance system causes the relevant bonds of the central methine atom to be considerably enlarged as expected, without however reaching the bond length of 1.550 Å observed in the unconjugated tetraphenylmethane [570]. The aromatic character of the rings proper is apparently modified in the direction

Table 11: Bond Lengths [Å] In Metal Chelates of Salicylaldehyde Imines (5.15)

No.	Substance	Ref.	*R*	*Me*	C_7-N	C_7-C_1	C_1-C_2	C_2-O	C_2-C_3	C_3-C_4	C_4-C_5	C_5-C_6	C_6-C_1	$N-R$
1	Salicylaldoxime	[539]	OH	—	1.270	1.459	1.411	1.358	1.389	1.371	1.393	1.379	1.391	1.405
2	Salicylaldehyde azine	[540]	*)	H	1.274	1.448	1.429	1.341	1.393	1.362	1.411	1.377	1.397	1.407
3	Bis(N-methyl-salicylaldiminato)-copper(II)	[541]	CH_3	Cu	1.285	1.454	1.420	1.312	1.418	1.366	1.392	1.365	1.420	1.510
4	Bis(N-methyl-salicylaldiminato)-zinc(II)	[542]	CH_3	Zn	1.312	1.414	1.415	1.323	1.420	1.396	1.41	1.37	1.43	1.47
5	Bis(N-methyl-salicylaldiminato)-dioxo-molybdenum(VI)	[543]	CH_3	Mo	1.288		1.381	1.348	1.420	1.385	1.388	1.382	1.405	1.475
6	Dichloro-N,N'-ethylenebis-[salicyl-aldiminato]-titanium(IV)-tetrahydrofuran	[544]	$-CH_2-$	Ti	1.285	1.446	1.401	1.384	1.388	1.386	1.384	1.363	1.431	1.469
7	N,N'-Ethylenebis-salicylaldiminato-pyridine-(vinyl)-cobaltate(III)	[545]	$-CH_2-$	Co	1.27	1.42	1.39	1.32	1.40	1.38	1.39	1.35	1.42	1.56
8	μ-N,N-Ethylenebis(salicylaldiminato) bis[2-methylallyl-palladium(II)]	[546]	$-CH_2-$	Pd	1.31	1.43	1.41	1.30	1.42	1.37	1.43	1.34	1.46	1.48
9	Bis(N-ethylsalicylaldiminato)-copper(II)	[547]	C_2H_5	Cu	1,30	1.43	1,405	1.335	1.395	1.355	1.355	1.345	1.415	1.47
10	Bis(N-isopropyl-salicylaldiminato)-copper(II)	[548]	C_3H_7	Cu	1.329	1.424	1.410	1.307	1.436	1.428	1.405	1.397	1.441	1.511
11	Bis(N-t-butylsalicylaldiminato)-palladium(II)	[549]	C_4H_9	Pd	1.287	1.473	1.408	1.308	1.421	1.391	1.401	1.390	1.428	1.511
12	Bis(N-phenyl-salicylaldiminato)-copper(II)	[550]	C_6H_5	Cu	1.29	1.42	1.39	1.28	1.41	1.35	1.35	1.37	1.41	1.45
	Averages				1.291	1.438	1.406	1.325	1.409	1.378	1.392	1.369	1.420	1.476
	Reference values of 9 three-dimensional structure analyses	[289]			1.295	1.430	1.415	1.312	1.415	1.388	1.384	1.367	1.423	

*) *R* = Salicylaldehyde imine

of a polyenic structure in consequence of the branching effect within the rings as opposed to tetraphenylmethane as reference substance. This is indicated in (5.17) by a partial fixation of the double bonds.

5.2. Conjugated Heterocycles

Having discussed the polymethine-aromatic intermediates of substituted benzenes we now turn to various heterocyclic compounds, of which only some display typical intermediates between polymethinic and aromatic structures. Frequently different meropolymethine structural groups which interact more or less strongly may be formulated in the molecule. In addition the usual presentation of the formulae requires in some molecules the formation of polyene-like structural groups which has in fact been observed. In consequence it must be taken into account that coupling effects may occur between the individual structural groups, which will be dealt with in Chapter 6. The consequences of a coupling effect for the light absorption of the heterocyclic compounds have not been studied so far. The changes in light absorption should be comparatively small as the interacting structural groups are, as a rule, only the first members of vinylene homologue polymethines or polyenes. It is for this reason that these compounds are represented in a separate chapter as has been the practice in the textbook literature. The substances are usually important in biology, so that extensive *X*-ray structural data are available. The heterocycles are another example of how molecular structures can be interpreted and their physico-chemical properties be estimated by means of the interaction of the three ideal states of unsaturated organic compounds.

A monomethine cyanine fragment competing with an aromatic π-electron sextet is to be found in the 2-amino-pyridine (5.18) [571a, b] and even more pronouncedly in the

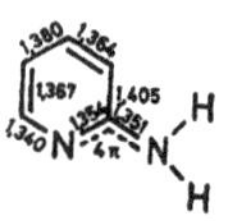

(5.18) [571a]

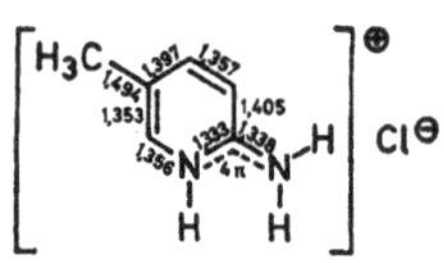

(5.19) [572]

(5.20) [70]

2-amino-5-methylpyridine hydrochloride (5.19). As in the substituted benzenes (Section 5.1), polymethine resonance induces a polyenic structure in the pyridine ring (5.20) which is aromatic when substituted.

In nearly the same manner an aza-trimethinemerocyanine fragment is realised in 2-amino-3-nitropyridine (5.21), whereas in the 3-aminopyridine (5.22) no indication of a polymethinic resonance can be pointed out within the pyridine ring in comparison to (5.21). But as in meta-substituted benzenes (Section 5.1) the C—N bond length of the amino group is slightly shortened.

Six symmetrically branched monomethine fragments are to be found in the cyanuric acid (5.23). The compound may also be interpreted by the superposition of several succinimide (4.7) or urea structures (4.11). The very accurately determined bond lengths coincide in fact with those of the succinimide. It would be interesting to know the influence on the structural parameters by ion-formation, but the structure of the cuprous salt [576] is too inaccurate for a discussion.

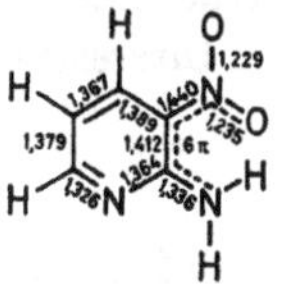

(5.21) [573]

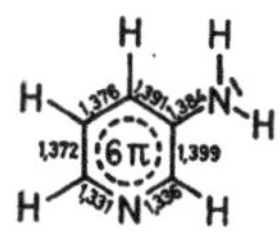

(5.22) [574]

In the barbituric acid (5.24) four monomethinemerocyanine structural groups can be formulated. Compared to (5.23) there is no resonance taking place at the CH_2 group. In consequence the two adjacent C=O bonds are shortened. Moreover, the shortening of the two C=N bonds in the upper half of the molecule seems to be significant. If,

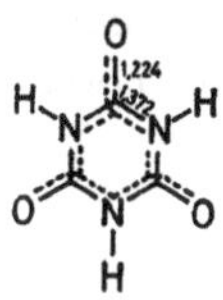

(5.23) [575]

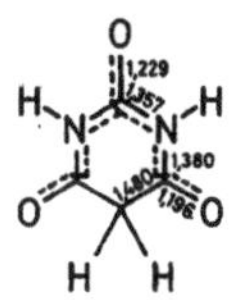

(5.24) [577]

however, an aromatic π-sextet is generated in the ring, as it happens in the barbiturate anion (5.25) or its 5-ethylderivative [579] the C—N bonds situated opposite will, on the average, be extended while the C—C single bond lengths will be reduced to an aromatic bond length of 1.40 Å. At the same time, the C=O double bonds obtain a larger single-bond character, owing to the aromatisation tendency of the ring.

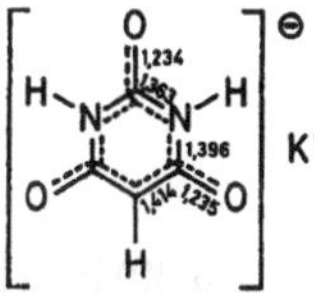

(5.25) [578]

Another explanation has to be given regarding the transformation from the 5,5-diethylbarbituric acid (5.26) [580, 581] and its derivatives [582—597] to the anion (5.27) [580, 598—601]. Since here, in contradiction to (5.24) and (5.25), the dialkyl-substituted group does not allow the formation of an aromatic structure, the de-

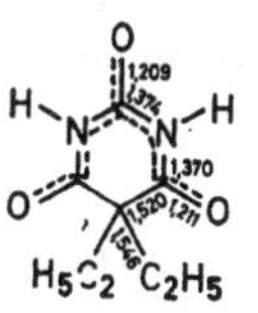

(5.26) [580]

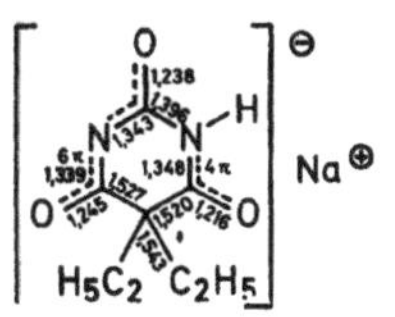

(5.27) [580]

protonation of a nitrogen atom will effect a strong resonance equalisation of the bonds within the succinimide structure, whereas the remaining monomethinemerocyanine structural group is clearly isolated by relatively long bond lengths. The effect is further intensified in the sodium-N-methyl-5,5-diethylbarbiturate [602, 603] or in the guanidinium 5,5-diethylbarbiturate dihydrate [604].

In view of the triad concept, the structure of the alloxan (5.28) is easy to interpret: The molecule consists of four monomethinemerocyanine structural groups. In consequence of branching of two of them the bonds involved are longer than the non-branched bonds. As is to be expected the C=O bond not involved in the polymethine resonance has the shortest bond length of the carbonyl groups. Similar structures are shown by alloxantin [606] or violuric acid monohydrate (5.29), for example. Interesting

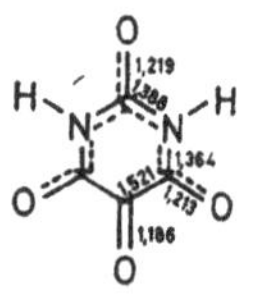

(5.28) [605]

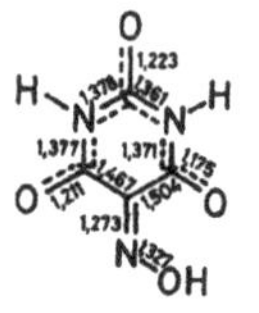

(5.29) [607]

is the influence produced on the structure of the latter compound by ion-formation in (5.30) [608, 609]. Additional resonance through a branched azatrimethine oxonol group is now possible, thereby shortening the C—C single bonds in the ring in analogy of (5.25), extending the exocyclic C=N bond and the two carbonyl groups and shortening the N—O bond accordingly.

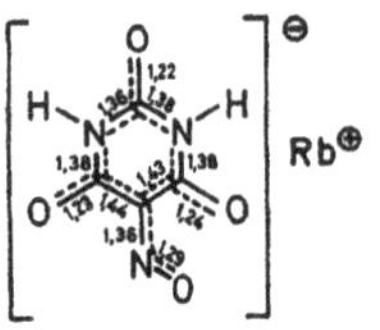

(5.30) [608]

In the uridines [610—613] and uracils (5.31) [614—623] as well as in the thymines (5.32) [624—626] and cytosines (5.33) [627—631] polyene-like structures occur in addition to the competing monomethine structural groups.

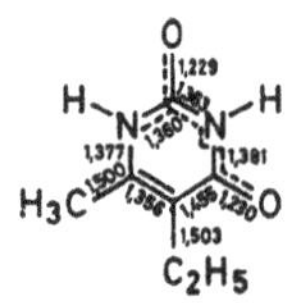

(5.31) [614]

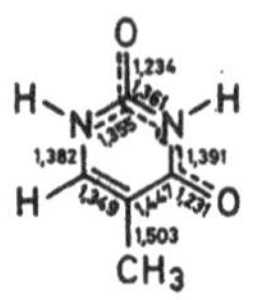

(5.32) [624]

The C=C double bond depicted is longer than the standard value of 1.33 Å to be expected (Table 1). Along with this, the lengths of the bonds adjoining the double bond are shorter than those of C_{sp^3}—C_{sp^3} single bonds, so that the polyene structure becomes more or less involved in the resonance of the total system, possibly accompanied

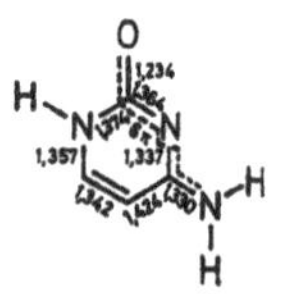

(5.33) [630]

by aromatic structural features. In (5.33) an azatrimethinemerocyanine fragment may be formulated in addition which helps to explain the relatively short C—N distance of the amino group.

Similar structural principles are to be assumed for the higher condensed heterocycles, where an isolation of individual structural groups becomes more and more difficult, owing to growing possibilities of resonance. The purine-like compounds such as the derivatives of guanine [632—637], adenine [638—643], xanthine [644, 645], hypoxanthine [646, 647], xanthazole [648], and uric acid [649, 650], of which the latter is

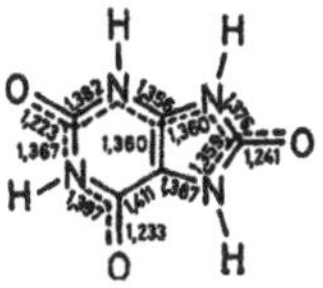

(5.34) [649]

given as an example in (5.34), are typical examples with two condensed rings. Common to most compounds is the relatively short, C=C bond joint to both rings which is part of a branched trimethinemerocyanine structural group. More structures built to a similar pattern are to be found in [651—659].

The coloured metal chelates of the 8-hydroxyquinoline (5.35), of which some more recently determined structures are given in Table 12, must also be included in the conjugated heterocycles. Here, same as in the meta-substituted benzene derivatives, only even-numbered polymethine radical structural groups, for example a dimethine chain between the N- and the O-atom, can be formulated.

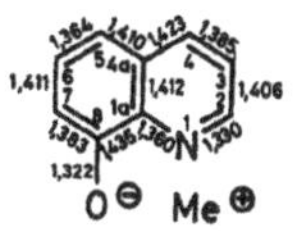

Average bond lengths taken from Table 12 (5.35)

The short C—O single bond of 1.32 Å indicates a polymethinic resonance. Accordingly the bond length is 1.35 Å when the oxygen is protonated and the free electron pair is not available on the quinoline nitrogen atom, as is the case in the 8-hydroxyquinoline-N-oxide [666]. In this context the large C—O bond length in the dichlorobis-8-quinolinolato-titanium(IV) [667] remains however unclear.

As in the quinolinium ions [668—670], a structure containing at the same time a polyene-like structural group is formed also in the 8-hydroxyquinolines, so that these compounds must be regarded as typical intermediates between all three main groups — aromatics, polyenes and polymethines. The characteristic lengthening of all bonds involved in the double branching of the ring systems is to be specially emphasised. As is to be expected, the 8-aminoquinoline-nickel-II chelate (Table 12, No. 5) possesses an analogous structure, whilst in a 4-amino-substituted quinoline (Table 12, No 6) there are indications of a normal trimethinecyanine chain between the two nitrogen atoms.

In luciferin [671, 672], the chromophore of the firefly responsible for chemiluminescence, the oxygen atom of the hydroxyl group is separated from the heterocyclic nitrogen atom by a tetramethine chain which leads to a preferentially aromatic structure with a C—O bond length of about 1.36 Å.

Three linearly condensed heterocyclic rings are contained in the coloured derivatives of flavine [673—691] and similar chromophores [692—694] which are of great interest

Table 12: Bond Lengths [Å] In Metal Chelates of 8-Hydroxyquinolines (5.35)

No.	Substance	*Me*	Ref.	$N-C_2$	C_2-C_3	C_3-C_4	C_4-C_{4a}	$C_{4a}-C_5$	C_5-C_6	C_6-C_7	C_7-C_8	C_8-C_{1a}	$C_{1a}-N$	$C_{1a}-C_{4a}$	C_8-O
1	Bis(2-methyl-8-quinolinolato) aluminium(III)-μ-oxo-bis-(2-methyl-8-quinolinolato) aluminium(III)	Al	[660]	1.335	1.423	1.374	1.417	1.417	1.367	1.428	1.377	1.429	1.359	1.402	1.333
2	Anhydrous Copper 8-hydroxy-quinolinate	Cu	[661]	1.324	1.403	1.335	1.415	1.399	1.363	1.435	1.375	1.428	1.355	1.417	1.321
3	Bis(8-hydroxy-quinoline)-silver(I)-pyridine solvate	Ag	[662]	1.334	1.404	1.344	1.440	1.417	1.368	1.394	1.396	1.436	1.383	1.408	1.320
4	8-Hydroxy-zinc-quinolinate	Zn	[663]	1.328	1.394	1.361	1.420	1.407	1.359	1.388	1.386	1.448	1.342	1.423	1.314
	Averages			1.330	1.406	1.358	1.423	1.410	1.364	1.411	1.383	1.435	1.360	1.412	1.322
															C_8-N
5	Nitrato-bis-(2-methyl-8-aminoquinoline)nickel(II)-nitrate	Ni	[664]	1.326	1.432	1.359	1.417	1.417	1.365	1.423	1.368	1.415	1.376	1.420	1.455
															C_4-N
6	7-Chloro-4-(4-diethyl-amino-1-methylbutyl-aminosulfate)-quinoline	—	[665]	1.317	1.377	1.387	1.468	1.400	1.379	1.408	1.357	1.420	1.373	1.405	1.314

as enzymes and coenzymes. The structure of the 5-acetyl-3,7,8,10-tetramethyl-1,5-dihydroalloxazine (5.36) shall serve as an example.

The structure of the aromatic π-sextet and, in analogy with alloxan (5.28), several monomethine structural groups stand out quite clearly. The comparatively short C=C

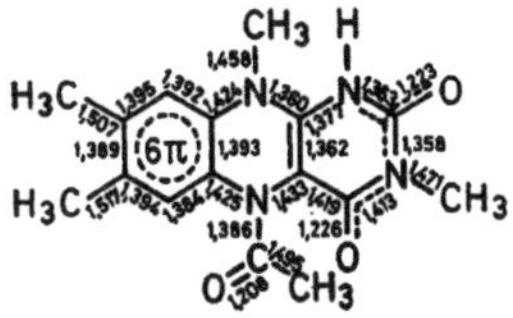

(5.36) [689]

bond between the two heterocyclic rings is part of potential trimethinemerocyanine structures (not depicted in formula (5.36)) between the two top nitrogen atoms and the carbonyl oxygen atom at the bottom of the right-hand ring. Owing to steric twisting of the acetyl group the monomethine resonance between the acetyl group and the ring nitrogen atom will be of little effect, as shown by the bond lengths.

6. Coupled Polymethines

In dye molecules containing more than one polymethine structural group with an odd number of chain atoms (2.1, $N = 3, 5, 7$) characteristic interaction effects will occur between the various structural fragments; those may be interpreted in terms of a coupling of two or more resonating systems. Many well-known dyes such as the quinone and indanthrene dyes and the indigoids are built on this principle. According to a proposal by Dähne [695a] dyes of that type are termed coupled polymethines. Similar ideas have been proposed more recently by Kiprianow [696]. In contrast to branched polymethines, where the characteristic π-electron density alternation is maintained through the branched chains, in coupled polymethines the charge alternation is perturbed, normally in such a way that two or more polymethine chains are linked together through equally charged atoms [695b]. A special case of coupling occurs in the indigoid dyes, in which polymethine chains are crossed by two or more common atoms in such a way that the charge density alternation of the one chain has opposite sign to that of the other chain causing a strong perturbation of the alternation [695b]. In normally coupled systems mentioned above the electronic structure of each coupling group, that is its bond lengths and electron densities, will change but little compared to the non-coupled systems, which was proved originally by Kulpe [152] and later on by several other authors by means of numerous X-ray analyses, described in the following sections, as well as by *NMR* and *IR* spectroscopic studies [697—699]. Light absorption [700—703], on the other hand, is substantially more influenced by the coupling effect. Bathochromic shifts of the absorption bands in the order of 200 to 300 nm are not unusual [695a, 702], with the bands showing strong solvatochromism [703] even in molecules of a highly symmetrical structure.

Quantum-chemical models explaining the coupling principle have been developed by Leupold et al. [355, 704] and recently, by Fabian et al. [705, 706]. The first mentioned succeeded in simulating the effect by empirically adjusting the coupling parameter. The latter subjected the electronic wavefunction to a configuration analysis. This way it is possible to relate the coupled systems to the participating subsystems and to rate the localisability of the polymethinic, polyenic and aromatic subsystems by indices.

Since all possible intermediates to polyenic and/or aromatic structures occur in the more complicated dyes of this type, we shall expound the fundamental relationships at first in simple molecules. Here it is possible to differentiate between various types of structure [695a, b]. In (6.1), for instance, two monomethine fragments are coupled by one carbon-carbon bond. In (6.2) two trimethine structural groups are linked to each other via bonds 1—2 and 4—5. In (6.3) one trimethine structural group interacts with a diene structural group, in a corresponding manner. Whereas at the beginning of this development the changes in light absorption were the only proof of coupling

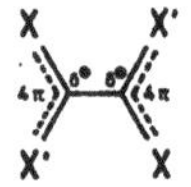
(6.1)

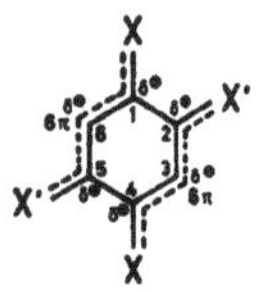
(6.2)
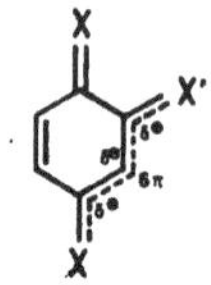
(6.3)

effects [355, 704], the *X*-ray structure analyses performed in the meantime revealed a new bonding property in coupled polymethines. From the quantum-chemical point of view it was expected that the bond lengths effecting the coupling were maximally 1.46···1.48 Å, e.g. the length of a single bond between sp^2 hybridised carbon atom (cf. Table 1). The coupling bonds however are unusually long. Bond lengths up to 1.56 Å have been observed. The most likely explanation of this effect given by KULPE [707] is an additional enlargement of the interatomic spacings by COULOMB forces which occur in the polymethine chains in consequence of the characteristic electron density alternation. As the carbon atoms adjacent to the coupling bonds are equally charged, a repulsion of the polymethinic fragments and hence a lengthening of the bonds is inevitable. In a more quantitative consideration this coupling effect caused by COULOMB forces presumedly will influence PAULING's rule of the invariance of the sum of bond orders and will have to be taken into account by an additional interaction term.

Depending on the symmetry of the π-electron distribution along the polymethine chains, all possible intermediates between ideal polymethines and ideal polyenes are conceivable, just as in the simple polymethines. In the following Sections we are going to apply these considerations to increasingly complicated compounds, though we will not be able to strictly keep to the systematics used in Chapters 3 and 4.

6.1. Ideal Polymethine-Polymethine Coupling

Coupled monomethine oxonols are realised in the dianions of oxalic acid (6.4) which have been already compiled in Table 3. The strong enlargement of the central $C_{sp^2}—C_{sp^2}$ bond, which other authors so far have been unable to explain, convincingly proves the coupling effect.

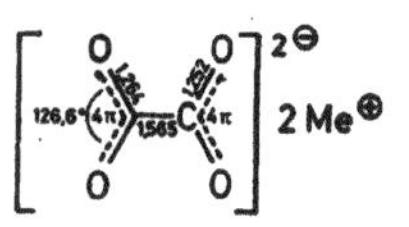
Average values taken from Nos. 5 and 6 of Table 3 (6.4)

A particularly suitable subject for studying the coupling effect is available when two symmetrical trimethineoxonols are linked to each other via two single bonds. This is the case in the red-coloured dianion of the 2,5-dihydroxy-benzoquinone-(1,4) (6.5) [708, 709]. As follows from *MO* calculations based on the coupling hypothesis [355, 705, 706], the π-bond orders within the polymethine chain should be largely equalised, whilst the two linking C—C bonds should have the smallest π-bond order and hence the largest bond length. By their trend, the *MO* results coincide with the *X*-ray data. Compared to the non-substituted 1,4-benzoquinone (6.6) [710—712], the [2,2] (2,5)-benzoquinophane

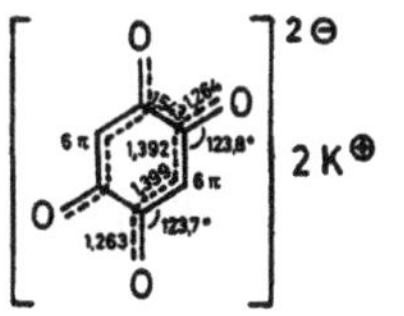
(6.5) [708]
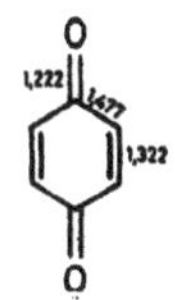
(6.6) [712]

[713a] or the 1,2-benzoquinone [713b], the C=O bond lengths also have been enlarged, from 1.222 Å to 1.26 Å, whereas at 1.543 Å the C—C bonds linking the oxonol fragments are enlarged in consequence of the Coulomb repulsion of the polymethinic fragments verifying the coupling effect.

Moreover, as in the simple polymethine chains (Section 3.1), the bond angles at the positively charged carbon atoms are significantly greater than 120°, due to the re-hybridisation effects in the π-electron framework [64]. In consequence the two adjacent oxygen atoms are bound to approach each other, causing in addition mutual repulsion of the two coupled groups.

The same structural principle of two coupled trimethineoxonols is to be found in the ammonium chloranilate monohydrate (6.7) studied by Anderson, where the C=O bond

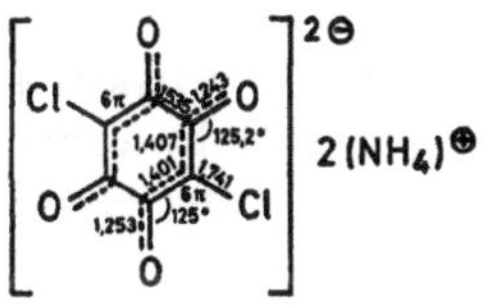
(6.7) [714]

lengths are somewhat shorter, possibly owing to the screening of the cation charges by water molecules. At 1.741 Å the Cl—C distance in (6.7) is greater than in the corresponding chloroquinones [715—720], e.g. in the 2,5-dichlorobenzoquinone-(1,4) (6.8).

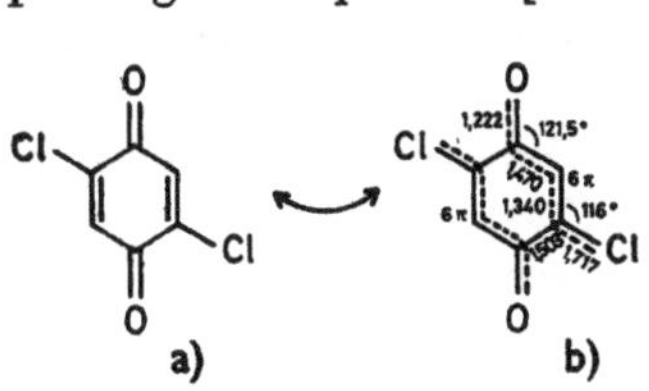
(6.8) [717]

In connection with the latter compound resonance participation of the chlorine atom in the opposite direction compared to (6.5) is discussed [716]. By comparison to quinone (6.6) the C=C double bond lengths in (6.8) are indeed larger while the C—C single bonds involved in resonance have shortened. Also the large C—C distances of 1.503 Å in the quinone ring seem to indicate a weak coupling effect of two trimethinemerocyanine structural groups (terminated by chlorine and oxygen) in accordance with the structure outlined in (6.8b). We consider these findings to be structure analytical evidence of the fact that atoms of group VII of the periodic table may act as terminal groups of polymethine systems in accordance with formula (2.1).

In the structure of the ammonium salt of the nitranilate (6.9) [721, 722] the C—C distances of 1.55 Å in the ring indicate an even stronger repulsion of the coupled structural groups. The C=O bond lengths, on the other hand, are short compared to (6.5) and (6.7). The effect may be attributed to a strong participation of the nitro groups in resonance, as indicated by the short N—C and comparatively long N=O bonds,

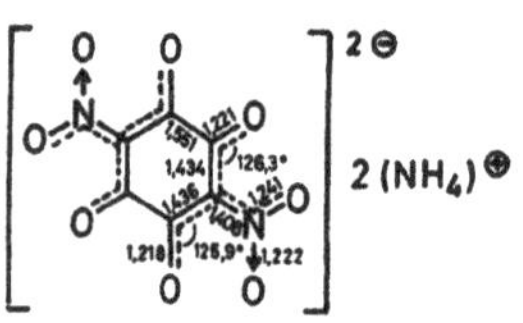

(6.9) [721]

in analogy to other polymethine-like nitro compounds (e.g. (3.32), (3.33), (3.35), (4.36) or (4.37)). As expected the branching effect of the two aza-trimethineoxonol structural groups, which is required for reasons of symmetry, leads to an extension of the C—C bonds in the ring to 1.435 Å against 1.40 Å in (6.5) or (6.7).

6.2. Coupled Meropolymethines In Simple Systems

It has been mentioned that with increasing asymmetry of the polymethine fragments in the coupled polymethines, the same polymethine-polyene intermediates will be observed as in the simple polymethine chains (Sections 4.1 to 4.4). For example, meropolymethine-like structures are obtained by protonation of the dianions of the coupled oxonols. The hydroxyquinone dyes, whose naphtho- and anthraquinone derivatives have a great importance, are of this type. Simple models are provided by oxalic acid (6.10), whose central C—C bond length at 1.51 Å is shorter than that of the

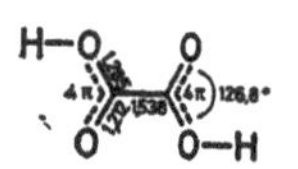

(6.10) [723]

coupled oxonols (6.4) but longer than that of a C_{sp^2}—C_{sp^2}-single bond (cf. Table 1), just as in the oxamates (cf. Table 3) which contain correspondingly coupled monomethinemerocyanine fragments.

Two merotrimethineoxonol fragments are coupled in the 2,5-dihydroxybenzoquinone-(1,4). For instance, the bond lengths of fluoranilic acid [724], and chloranilic acid (6.11) as well as its dihydrate [726] are intermediate between those of the dianion

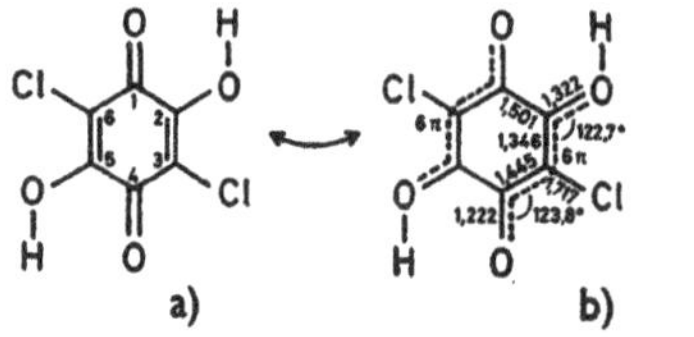

(6.11) [725]

(6.7) and those of benzoquinone (6.6). The bonds C_1-C_2 and C_4-C_5 of 1.501 Å which cause the coupling effect are significantly longer than in (6.6) and shorter than in (6.7). The same holds for the double bonds C_2-C_3 and C_5-C_6. The C_3-C_4 and C_6-C_1 single bonds within the merooxonol chains are appreciably shortened, and so are the C—O single bonds (cf. Table 1), suggesting a resonance in accordance with structure (6.11b). In agreement with this interpretation, the bond angles at the positively charged methine atoms lie between 120° and 126°. Here as in the non-coupled merooxonols (Sections 4.2.3, 4.3 and 4.4) the carbonyl double bond is hardly affected. As shown by the structural data, the conjugation along the unsymmetrical oxonol chains predominates at the expense of a likewise conceivable conjugation between the chlorine and carbonyl substituents, as observed in (6.8). The O-methylobtusaquinone [727a] which contains two methoxy groups may be accommodated within the same category. Also the tetrahydroxy-benzoquinone-(1,4) [727b] should be mentioned here. For reasons of symmetry, however, its structure resembles more the benzoquinone itself.

Two unsymmetrical trimethineoxonol fragments are also contained in the nitranilic acid, whose structure (6.12) was determined in the form of the hexahydrate [728, 729]. By analogy with the dianion (6.9), the nitro groups participate in the mesomeric system,

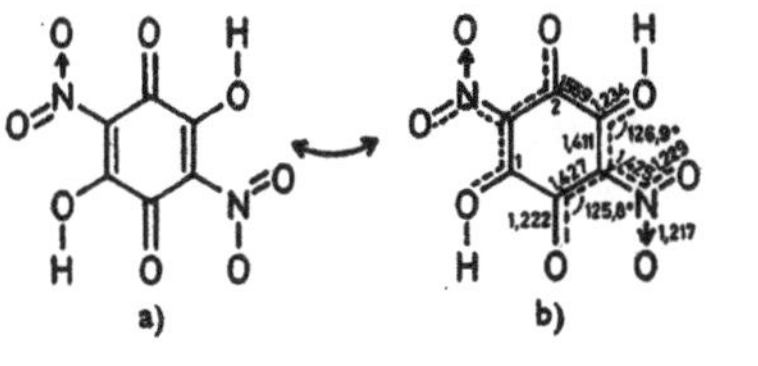

(6.12) [728]

even though the shortening of the C—N distance is less pronounced. Compared to (6.11) the bonds have undergone distinct changes. The alternation of the bond lengths within the meropolymethine chain disappears almost completely. Particularly striking is the shortening of the C—O single bonds to a distance comparable to the carbonyl double bond lengths, thereby enabling the compound to approach a highly symmetrical structure in accordance with formula (6.12b). There is an enlargement both of the carbon-carbon single bonds, which effect the coupling, up to 1,56 Å and of the adjacent bond angles of the oxonol fragments up to 127°. In consequence of this bonding type the basicity of nitranilic acid is extremely low, thus enabling the protons to form hydronium ions with the water molecules of the crystal lattice [728]. As expected, the same structural principle is realised in the hexahydrate of cyanilic acid [729]. The bond lengths are similar to those in (6.12).

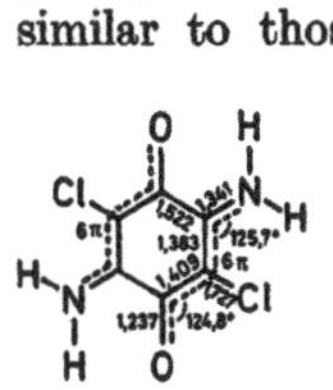

(6.13) [152]

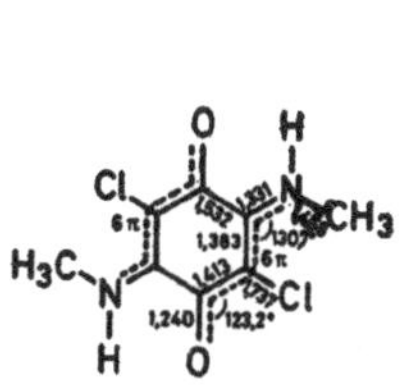

(6.14) [108]

Coupled trimethinemerocyanine structural groups are included in the diaminobenzoquinones, which are also of red colour [701, 704]. The main points of the coupling concept were confirmed experimentally in 1966 for the first time by X-ray analysis [730]. The structures of 3,6-dichloro-2,5-diaminobenzoquinone-(1,4) (6.13), 3,6-dichloro-

2,5-bis-[monomethylamino]-benzoquinone-(1,4) (6.14), and 2,5-bis(hydroxyethylamino)-benzoquinone-(1,4) (6.15) are identical within the significance of 2σ.

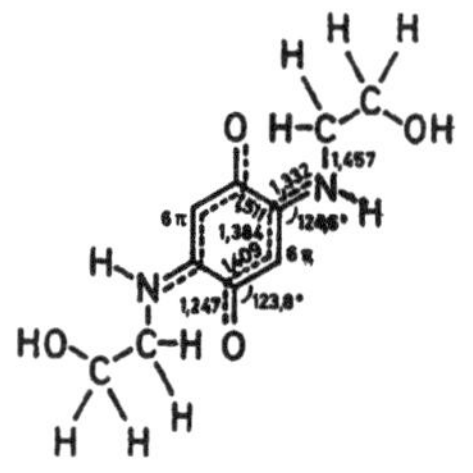

(6.15) [731]

Compared to the benzoquinone (6.6), or its dichloro derivative (6.8), the C—C-double- and single-bonds are clearly equalised within the merocyanine chains, though a small bond length alternation remains. The C—N and C=O bond lengths are identical with the distances in the non-coupled merocyanines (Sections 4.2.1, 4.3 and 4.4). Here, as well, the bonds which cause the coupling effect are unusually long and the bond angles at the positively charged methine atoms are enlarged. Alkyl substitution in (6.14) and (6.15) seems to strengthen the effect in that — within the significance point though — the C—N bonds are further reduced while the coupling bonds at least in (6.14) are still further extended.

The 2,5-bis-[pentamethylene-imino]-benzoquinone-(1,4) (6.16) [106, 107, 732] and the 2,5-bis-[ethylene-imino]-benzoquinone-(1,4) (6.17) show distinct differences. In the first-mentioned compound the amino groups have been twisted out of the molecular

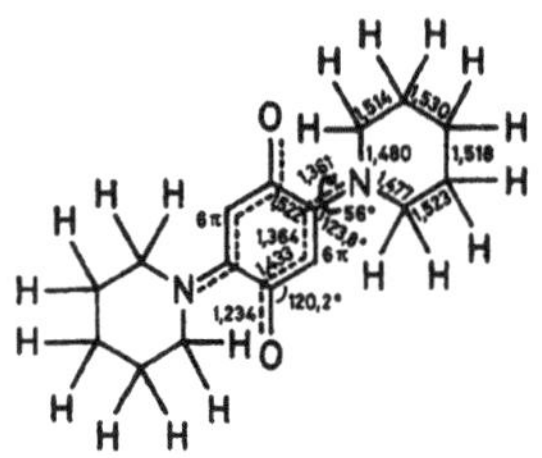

(6.16) [106]

plane of the quinone ring, thereby diminishing the resonance within the trimethine chains, which gives rise to a stronger alternation of bond lengths inside the trimethine chains compared to (6.13), (6.14) and (6.15). The structure of the ethylene-iminoquinone

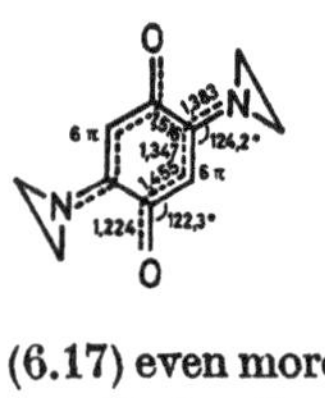

(6.17) [733]

(6.17) even more closely approximates the benzoquinone structure. This is not surprising considering the very weak basicity of the ethylene-imino group and explains why compounds of this kind have been treated as substituted quinones in the light of the discussion of spectroscopic data [354, 734]. The colour of the substances decreases as the quinone character becomesstronger [695a, 734].

Coupling effects between one trimethine and two monomethinemerocyanine structural groups are likely to determine the structure of the 4-chloro-5-methylamino-2,3,6-pyridine trione (6.18). The trimethine structure clearly stands out by its equalisation of the C=C bond lengths, the short C—N bond and the large bond angle at the positively charged carbon atoms. The monomethine structures with one common nitrogen atom correspond by their distances to the succinimide structure given in Section 4.2.1.

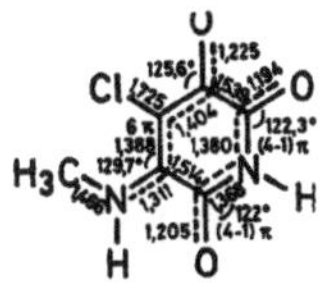

(6.18) [735]

An even more complicated structure is to be found in the aminoazimines, of which the triazolidine (6.19) was subjected to *X*-ray analysis. The authors [736] pointed out the isolation of two partial chromophores. We are obviously dealing here with a coupling effect between a neutral monomethinemerocyanine fragment and two joint monomethine structural groups of opposite charge whose complex electronic structure needs to be studied in greater detail. It may be that the Coulomb contraction between the two charged fragments causes the relatively short N—N bond of 1.36 Å.

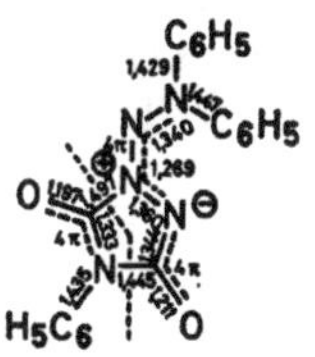

(6.19) [736]

In both the coupled trimethinemerocyanine and in open-chain merocyanines (Section 4.3) the C=O bond lengths are shorter than originally estimated from *MO* calculations on the basis of the polymethine concept. This phenomenon which is characteristic of merocyanines is studied in detail in [353, 356].

In the crystal lattice the molecules of the aminoquinones are remarkably close-packed [108]. Structure analyses have revealed the same pattern of packing as mentioned earlier in relation to the simple polymethines (Section 3.1), which brings forth the shortest intermolecular distances between the positively and the negatively charged atoms fitting the charge density alternation in the chains.

6.3. Polyene-Polymethine and Aromatic- Polymethine Coupling

The deep colour and the solvatochromism of the monoaminoquinones (6.20) were the reasons for postulating, in analogy to the concept of coupled polymethines, a coupling between polyene and polymethine structural groups [695a, 698, 700, 703].

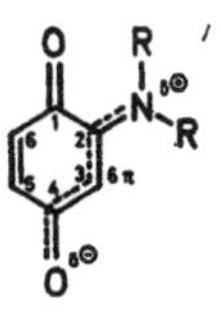

(6.20)

It was possible to indirectly confirm this prediction by *IR* and *NMR* spectroscopic investigations [698, 700]. A first indication that the polyene-polymethine coupling is also reflected in structure analyses was provided by the *X*-ray structure of α-5-(2'-chloroethoxy)-1,2-benzoquinone-2-oxime (6.21), where the double and single bonds on the "polyene part" manifest a somewhat stronger alternation than those on the "polymethine part". A similar effect should however occur also in the mono-hydroxy-benzoquinone(-1,4) (6.22) which, for reasons not yet understood, is not the

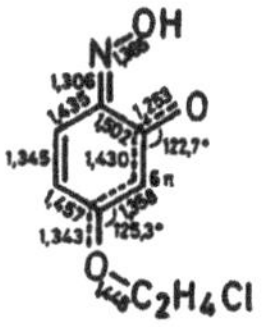

(6.21) [737]

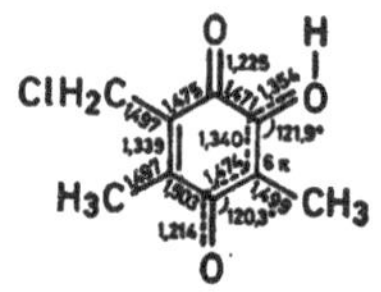

(6.22) [738]

case, with the exception of the relatively short C—OH bond and the long $C_4—C_5$ ring bond. In contrast 2-methyl-5,6-dimethoxybenzoquinone-(1,4) shows the expected trend towards a one-sided extension of the quinone-like C=C double bond on the methoxy-substituent side from 1.322 Å to 1.368 Å [739]. Other compounds containing polyene-polymethine structural groups have been dealt with in Section 5.2.

In 1-amino-2(para-chlorophenyl)-indenone-3 (6.23) there are indications of a coupling effect between a trimethinemerocyanine structural group and an aromatic π-electron sextet. The bond lengths in the aromatic ring on the left side average 1.40 Å. The tri-

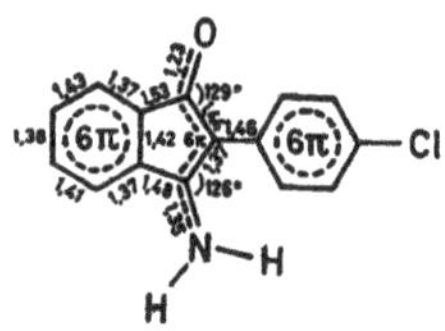

(6.23) [740]

methinemerocyanine structural group possesses similar bond lengths as the trimethine chains in (6.16), the bond angles are widened at the positively charged methine atoms, and the coupling bonds are correspondingly long with 1.53 and 1.48 Å, respectively. There seems to be a direct relation between the enlargement of the angles and of

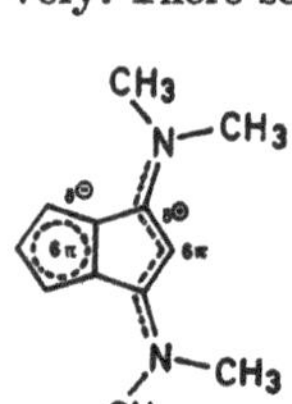

(6.24)

the coupling bonds. The greater the angle, the higher is the positive charge at the methine atom and thus the stronger are the COULOMB repulsion forces, despite the fact that there should be no appreciable charges at the carbon atoms of the aromatic ring. Other substances like the 1,3-bis-[1'-alkylpyridinium]-cyclopentadienides [939] and pentalene (6.24) can be stabilised in the same way [741]; no *X*-ray analysis of the latter compound is available, however.

6.4. Coupling Effects In Naphthoquinone Dyes

The structure of the naphthoquinone derivatives is modified by the aromatisation tendency of the condensed benzene ring and by possible branching effects between the two rings. For comparison, the accurately determined bond lengths of 2,2'-dinaphthoquinone-(1,4) are presented in (6.25), which are equivalent to those of the 1,4-naphthoquinone [743] and its derivatives [744]. The formation of a double bond

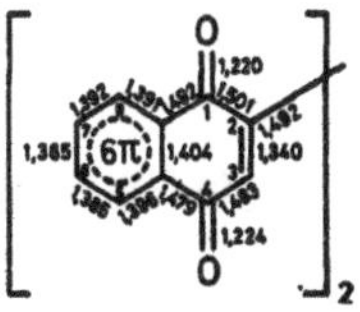

(6.25) [742]

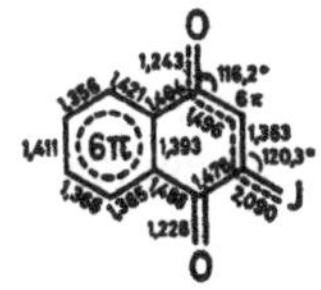

(6.26) [745]

in the central part of the molecule is obviously prevented by the cyclic 6π electron system with equal bond lengths, corresponding to those of benzene, causing the adjacent single bonds of the quinone ring to be extended to a greater degree than in the benzoquinone (6.6).

The effect produced by halogen substitution of the 1,4-naphthoquinone in the 3-position is the same as in the halogen-substituted benzoquinones (6.8): equalisation of bond lengths between the carbonyl oxygen atom and the halogen atom in 3-position, owing to the formation of a meropolymethine-like structure, as demonstrated by the example of the structure of 3-iodine-naphthoquinone-(1,4) (6.26). Just as in (6.8) the influence of the polymethinic charge alternation on the bond angles is small or even of opposite direction.

It appears that the coupling effect induces in addition a polyenic bond alternation in the condensed benzene ring. Unfortunately the structures of the 3-bromo- and 3-hydroxynaphthoquinone [746–748] fail to be accurate enough for verifying these statements. In this context, the structure determinations announced [749] of the 3-halogen-naphthoquinone-(1,2) could be of interest.

In (6.27) a competition between an iodine-terminated trimethine structure and an unsymmetrical oxonol structure is conceivable. As in (6.11) the resonance within the oxonol structure predominates. Striking is the very short C—OH bond of 1.307 Å compared to the hydroxybenzoquinones (Section 6.3).

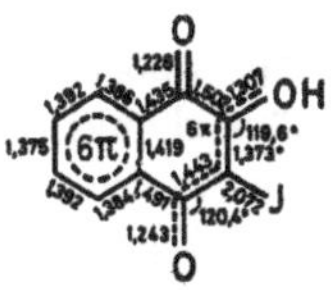

(6.27) [750]

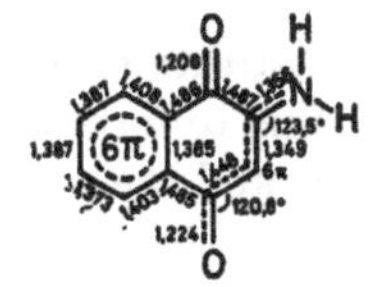

(6.28) [751]

Further evidence of the coupling concept is provided by the 2-aminonaphthoquinone-(1,4) [751–754] and the 4-aminonaphthoquinone-(1,2) [755–757] which is proved, for instance, by the structures (6.28) and (6.29). The trimethinemerocyanine structure in the para-quinone (6.28) is easily detectable, though it shows stronger polyene features than in the corresponding aminobenzoquinones (6.13) to (6.16).

Contrary to this behaviour, the bond lengths within the trimethine chain of the ortho-quinone (6.29) are nearly equalised, with the C—N distance appreciably shortened and the C=O distance enlarged forming a nearly ideal polymethine structure with symmetrical π-electron distribution. On halogen substitution in the ortho position relative to the amino group the resonance within the trimethinemerocyanine group will dominate, as in the aminobenzoquinones (6.13) and (6.14), compared to the likewise possible conjugation within a trimethine chain with a terminal halogen atom. When the conjugation of the amino group is interrupted by a methylene group, as is the case in the 3-N-methylanilinomethylnaphthoquinone-(1,4), a nearly ideal naphthoquinone structure is again obtained [744].

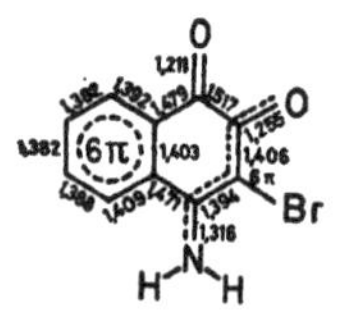

(6.29) [755]

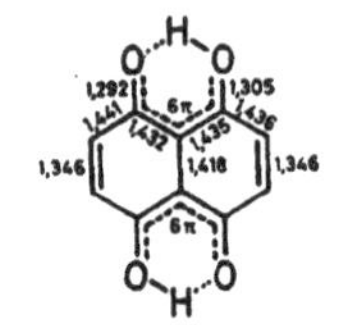

(6.30) [758]

A polyene-polymethine coupling effect is to be expected in the naphthazarin (6.30) which should be intensified by forming the symmetrical dianion. To date X-ray analyses are only available of the protonated molecule which exists in diverse modifications [758—761]. However, the average bond lengths of two accurate structure analyses presented in (6.30) support the predicted structural principle with the long C=O distances of the trimethineoxonol fragments being particularly convincing.

6.5. Coupling Effects In Anthraquinone Dyes and More Complex Systems

The structure of the anthraquinone dyes should be seen in comparison to the structure of the anthraquinone (6.31) [762, 763], which is determined, to an even greater extent than the naphthoquinone (6.25), by the tendency of the condensed benzene rings toward aromatisation. The 10,10′-dianthronyl [764] presents a similar example.

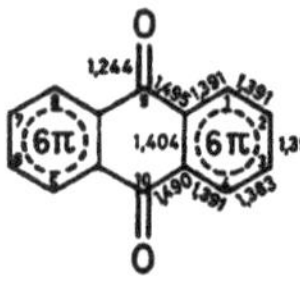

(6.31) [762]

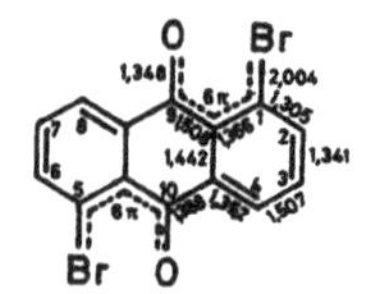

(6.32) [765]

In the case of halogen substitution of anthraquinone in the 1,5 positions a polyene-polymethine coupling according to formula (6.32) may be anticipated. This should be modified by branching effects and the aromatisation tendency of the cyclic π-electron sextets. However, the available less accurate structure analyses [765, 766, 767] do not admit of any structural interpretations.

The coupled structure indicated in (6.32) should be further strengthened by introducing in the 1,5 position substituents with stronger donor properties such as hydroxy or amino groups. Some examples of X-ray analyses are given in (6.33) to (6.35). Here again, the accuracy of the structure analyses available thus far does not suffice to make definite statements. As expected the bond lengths of the hydroxyl and amino groups in the direction to the quinone ring are generally somewhat reduced in favour of a slight enlargement of the carbonyl bond. On the other hand, there seem to be difficulties in the formation of polyene-like double bonds in analogy to structure (6.32). In the majority of the structures available a preferential fixation of a double bond is

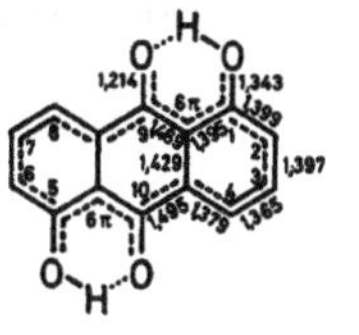

(6.33) [768, 769]

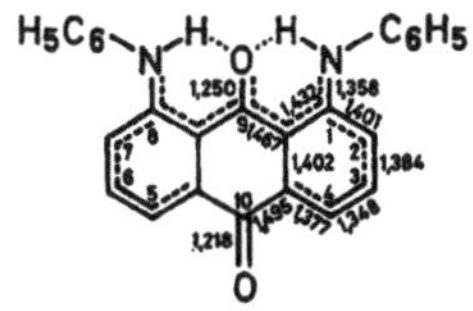

(6.34) [770]

to be found in the 3,4 and 5,6 as well as in the 7,8 position. It seems that with these more complicated dyes we have reached the limit where application of the coupling concept is possible, because of the competition between polymethinic, polyenic and aromatic structural groups.

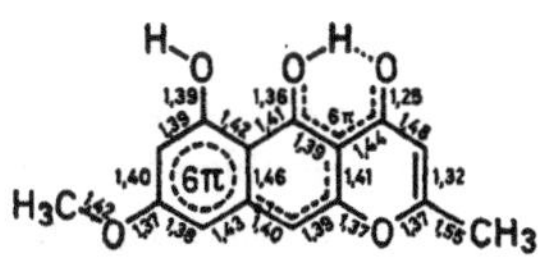

(6.35) [771]

More structure analyses of hydroxy compounds are available: chrysazin [772], anthraquinonethiadiazole [773], celebixanthone [774], 2-hydroxy-1,2,4,7-tetramethoxyxanthone [775], γ-rhodomycinone [776] and 1,5-dinitro-4,8-dihydroxyanthraquinone [777]. The structures of corresponding amino derivatives also have been determined: N,N'-diphenyl-1,5-diaminoanthraquinone [778] and 4,4-diamino-1,1'-bianthraquinone [779].

At present even greater uncertainties exist in interpreting more complicated dyes, such as the indanthrone (6.36), the hydrazone derivatives of the 2,2'-di-(1,4-naphthoquinone) [781], or the γ-quinacridone [782], whose structure analysis has been announced. Nevertheless there appears to be a certain trend in the indanthrone (6.36)

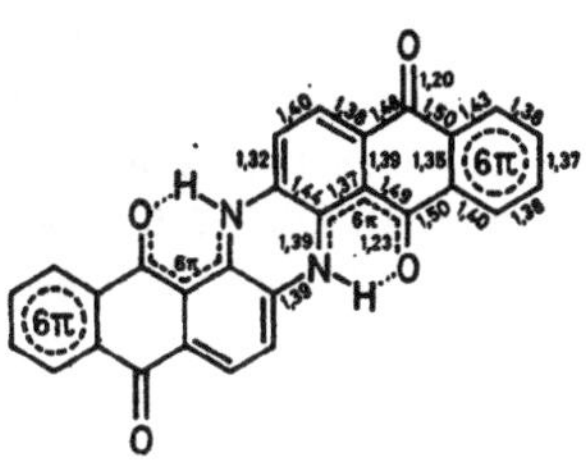

(6.36) [780]

towards the formation of trimethinemerocyanine structural groups, indicated by the slight enlargement of the carbonyl distances and the shortening of the C—N bonds. On the other hand, the aromatic π-sextets with a relatively high degree of bond length equalisation and the remaining dienone structural groups with alternating bond lengths are fairly outlined.

6.6. Indigoid Dyes

For over a hundred years, indigo (6.37), the "king of dyes", has fascinated dye chemists by its deep colour and high stability [8, 695a, 783]. The ability of a colour theory to interpret the indigo structure even seems to be a measure of its usefulness.

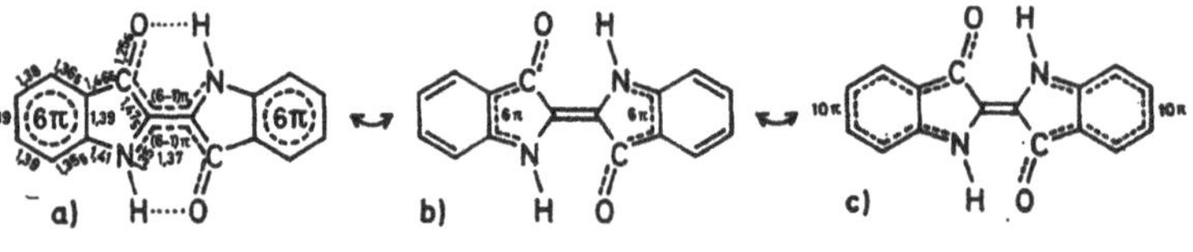

Averages taken from [792, 793] (6.37)

The two crossed trimethine fragments at the central part of this and similar molecules are occupied by (6—1)π electrons only, since two of the π-electrons are common to both fragments.

Dähne and Leupold attributed the colour of the indigo to the strong coupling and intramolecular interaction of several merocyanine structural groups and two aromatic π-sextets in the molecule [355, 695a]. On the one hand, it is possible to formulate via the central C—C bond two crossed trimethinemerocyanine structural groups (6.37a) in which the charge alternation is completely perturbed at the common crossing atoms. The crossed trimethine fragments are in resonance with two trimethinemerocyanine structural groups by way of the two outer ring carbon-atoms (6.37b). On the other hand, we have to take into account, apart from the aromatic π-electron sextet structure in (6.37a), the mesomeric involvement of two heptamethine structural groups around the benzene rings (6.37c) on the grounds of the typical substituent

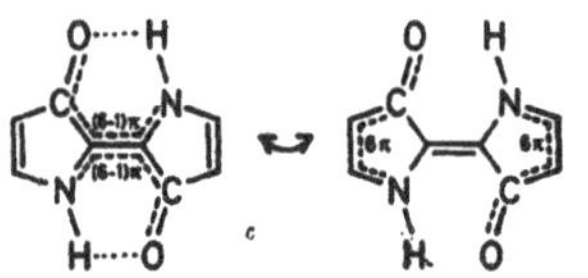

(6.38)

effects in the condensed benzene rings [784, 785]. Provided this hypothesis is justified, the polyenic character in the merocyanine fragments ought to increase as the possibilities for resonance in the indigo decrease, that is as the conjugated system is reduced. This problem has been studied in the last few years by Lüttke et al. [783, 786—790]. The authors were able to show that the so-called H-chromophore (6.39) is to be considered the smallest structural unit resembling certain indigo properties. Indeed, the colour of the indigo-like compounds diminishes in the order (6.37) > (6.38) > (6.39). Thus far no reliable, comparable X-ray analyses are available of the three types of

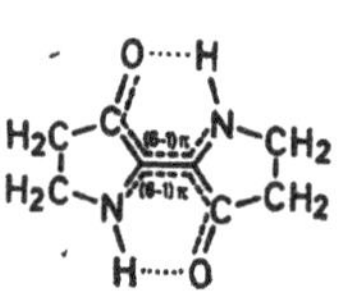

(6.39)

compound. But infrared spectroscopic data [787, 789] and quantum-chemical calculations [790, 791] prove the polyenic character (shortening of the C=O bond and enlargement of the C—N bond in the H-chromophore (6.39)) to be more pronounced than in indigo (6.37). However, with regard to the C=O bond length the pyrrole indigo (6.38) seems to be an exception insofar as the theoretical C=O bond order turns out to be smaller than the bond order of indigo (6.37). Experimentally, there results in the indigo chromophore an increase in the C=O stretching frequencies in the order (6.37) < (6.38) < (6.39) [787]. In the thioindigo series the same frequencies correlate with quantum-chemical data [789].

The argumentation concerning the interpretation of the indigo structure is going on. In a more recent work by FABIAN et al. [706], a configurational analysis of the pyrrole indigo chromophore was carried out based on PARISER-PARR-POPLE-SC wavefunctions which showed that there is no point in studying individual, isolated polymethine fragments in the indigo because of the strong coupling present already in the ground state. But this still does not explain the intense colour of the indigoids. After all, the 14-π electron system of the pyrrole indigo chromophore absorbs at wavelengths longer by several hundred nanometres than in aromatic compounds of comparable size (e.g. phenanthrene, anthracene). Hence the extent is to be studied to which more general polymethine features, which do not depend on a particular molecular structure, such as perturbation of the charge alternation in the ground state and a pronounced charge redistribution on light absorption, will be detectable also in strongly coupled systems. We think this will be possible so that it is justified, in spite of the originally voiced doubts [354, 791], to give a qualitative interpretation of the indigo structure based on a strong resonance between different, coupled polymethine structural fragments.

The two *X*-ray analyses available of indigo to date are not very accurate [792, 793], but they have a number of structural elements in common which may corroborate the coupling hypothesis, such as shortening of the C—N bond and enlargement of the C—O and the central C = C bonds. The average values from the two structures are given in (6.37a). The structures of the thioindigo (6.40), selenium indigo [795] and oxin-

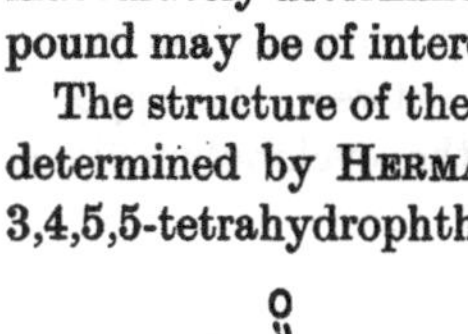

(6.40)

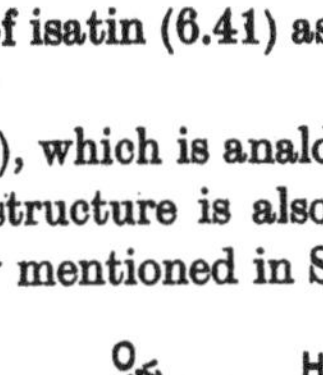

(6.41) [797]

Averages taken from [792, 794]

digo [796], which correspond to indigo, are useful for comparison, but even more inaccurately determined. Also the structure of isatin (6.41) as a colourless reference compound may be of interest.

The structure of the H-chromophore (6.42), which is analogous to the thioindigo, was determined by HERMANN et al. A similar structure is also realised in the 3,6-dithia-3,4,5,5-tetrahydrophthalimide (6.43) already mentioned in Section 4.2.1.

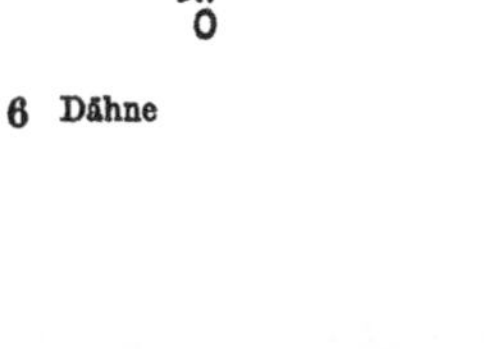

(6.42) [798]

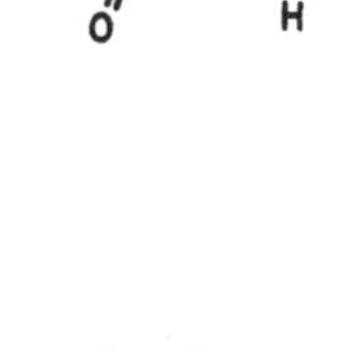

(6.43) [374]

The structural data available for the colour-theoretically interesting structures of the isoindigo (6.44) and the indirubin (6.45) are also not very accurate; in (6.44) the polyene character through the central C=C bond is more developed than in the indigo

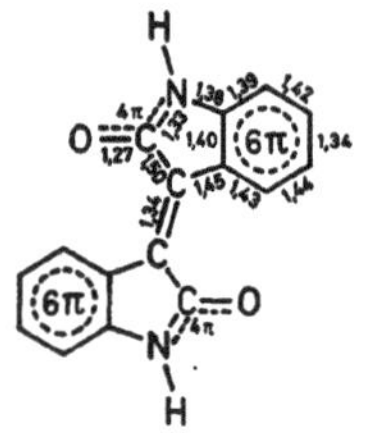

(6.44) [799]

or indirubin, whilst the latter is noteworthy by its unequal carbonyl bond lengths. This indicates a more pronounced polymethine character in the lower part of the molecule shown in formula (6.45).

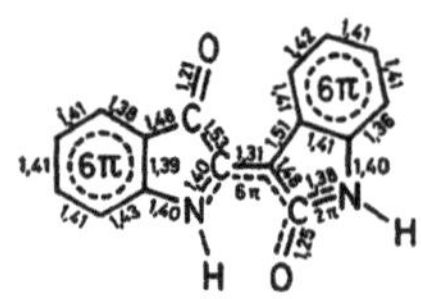

(6.45) [800]

Related to indigo is the basic chromophore of the pigments of red hair [801, 802], the trichochromone (6.46), which contains two crossed trimethinethiacyanine chains. The structure of the diphenyl derivative (6.46), R = p-C_6H_4-Br was determined for two red cis-isomers and the yellow trans-isomer. Formula (6.46) shows the average bond

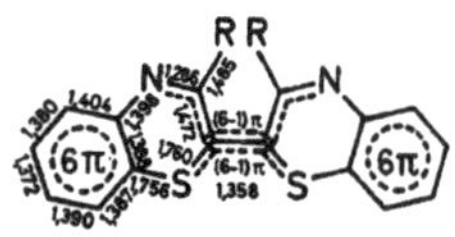

(6.46) [802]

lengths for the more stable cis-isomers. The bond lengths are not very informative because of the comparatively strong mutual steric hindrance of the two ring systems in the molecule.

6.7. Cyclically Coupled Polymethines

A particularly strong resonance between individual polymethine structural groups is to be expected in the squarate and the croconate dianions. In the squarate ion (6.47), two crossed and double branched trimethine-oxonol fragments between the opposite situated oxygen atoms can be formulated. The structures of the cyclobutenediones

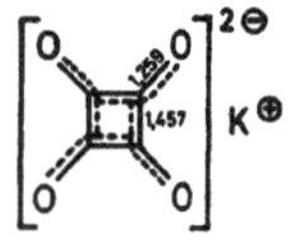

(6.47) [803]

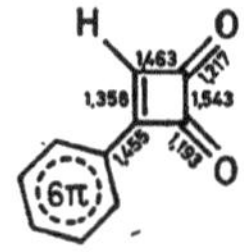

(6.48) [804]

[804, 805] may be used for comparison, for example the phenylcyclobutenedione (6.48), in which diminished conjugation leads to shorter C=O bond lengths and larger C—C bond lengths in the four-membered ring, with the exception of the shortened C=C double bond.

With symmetrically branched trimethinemerocyanine structural groups the 2,4-bis-(diethylamino)-cyclobutadiene-1,3-dicarboxylic acid diethyl ester (6.49) may be regarded

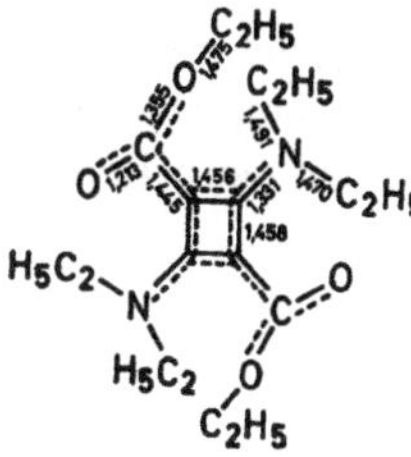

(6.49) [806]

as a derivative of the squarate ion. But whereas in the first substance the charge alternation is partly compensated because of crossing of several polymethine fragments, in (6.49) a pronounced charge alternation will occur; thus it is questionable if this compound can be systematised as a coupled polymethine. A striking feature of this compound is the shortness of the C—N bond lengths, corresponding at 1.33 Å to those of the cyanine cations (Section 3.1). The relatively short C—C bond lengths of the four membered ring indicate strong resonance in comparison to other conjugated cyclobutenes with C—C distances of 1.5 to 1.6 Å (cf. (6.48) and [807, 808]).

Various salts of the croconate ion (6.50) have been investigated, the structural data of which are summarised in Table 13. Here an unsymmetrical structure is clearly indicated by two strongly shortened C=O bond lengths. As a consequence of this

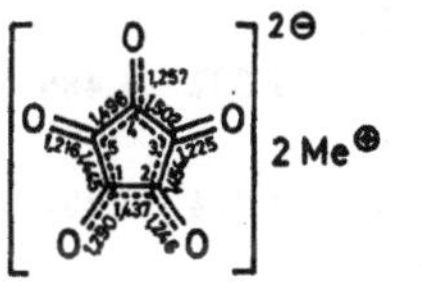

Average bond lengths taken from Table 13 (6.50)

Table 13: Bond Lengths [Å] In Crononate Ions (6.50)

No.	Cation	Ref.	C_1-C_2	C_2-C_3	C_3-C_4	C_4-C_5	C_5-C_1	C_1-O	C_2-O	C_3-O	C_4-O	C_5-O
1	$(NH_4^+)_2$	[809, 810]	1.452	1.463	1.455	1.463	1.452	1.289	1.241	1.269	1.269	1.241
2	NH_4^+, H^+	[810]	1.42	1.44	1.55	1.56	1.40	1.37	1.18	1.19	1.24	1.21
3	Rb^+, H^+	[810]	1.45	1.47	1.54	1.54	1.40	1.34	1.27	1.21	1.25	1.23
4	Cu^{++}	[811]	1.446	1.432	1.491	1.459	1.457	1.244	1.276	1.243	1.240	1.249
5	$Zn^{++}\cdot 3H_2O$	[811]	1.415	1.475	1.501	1.484	1.480	1.247	1.254	1.217	1.274	1.186
6	$Mn^{++}\cdot 3H_2O$	[812]	1.442	1.445	1.473	1.472	1.482	1.252	1.259	1.222	1.269	1.180
	Averages		1.437	1.454	1.502	1.496	1.445	1.290	1.246	1.225	1.257	1.216

shortening, one each of the adjacent C=C bonds is enlarged. The remaining bond lengths point to a preference of the branched trimethineoxonol structures $C_2-C_3-C_4$ and $C_1-C_5-C_4$, which are indicated in (6.50).

The chromophores of the porphines (6.51) and related compounds have been also considered as cyclically coupled polymethines with two times 9π-electrons occupying two times 8 chain-atoms [395a]. We will follow this classification despite the fact that there is no perturbation of the π-electron density alternation through the ring. Obviously this alternation facilitates the cyclic conjugations, which gives rise to an abnormally high diamagnetism of these compounds [813, 814]. The fact that here preference is given to the "internal" conjugation illustrated in (6.51) has recently been re-confirmed [815]. More structural data are summarised in Table 14.

(6.51)

Table 14: Average Bond Lengths [Å] In Substituted Porphines (6.52)

No.	Ligand	*Me*	*R*	Ref.	$N-C_1$ ($N-C_4$)	C_4-C_5 (C_5-C_1)	C_1-C_2 (C_3-C_4)	C_2-C_3	C_5-R
1	Porphine	$2H^+$	H	[816]	1.366	1.386	1.442	1.342	–
2	Tetraphenylporphine	$2H^+$	C_6H_5	[817]	1.351 (1.349)	1.411 (1.394)	1.444 (1.432)	1.362	1.514
3	Tetraphenylporphine	Cu^{++}	C_6H_5	[818]	1.383 (1.386)	1.380 (1.358)	1.461 (1.436)	1.337	1.486
4	Tetraphenylporphine	Zn^{++} $(H_2O)_2$	C_6H_5	[818]	1.369 (1.379)	1.414 (1.403)	1.421 (1.429)	1.374	1.497
5	Tetraphenylporphine	$SnCl_2^{++}$	C_6H_5	[819, 820]	1.370	1.407	1.446	1.380	1.494
6	Tetraphenylporphine	Pd^{++}	C_6H_5	[818]	1.369 (1.377)	1.413 (1.386)	1.440 (1.432)	1.346	1.502
7	Tetraphenylporphine	$(CoNO)^{++}$	C_6H_5	[821]	1.378 (1.374)	1.391 (1.396)	1.429 (1.431)	1.360	1.500
8	Bis(piperidine)-tetraphenylporphine	Co^{+++}	C_6H_5	[822]	1.383 (1.385)	1.386 (1.391)	1.435 (1.435)	1.356	1.494
9	Tetraphenylporphine	$(CoCe)^{++}$	C_6H_5	[823]	1.386	1.381	1.460	1.368	1.529
10	Bis(piperidine)-tetraphenylporphine	Fe^{++}	C_6H_5	[824]	1.384 (1.384)	1.396 (1.395)	1.443 (1.445)	1.347	1.501

No.	Ligand	*Me*	*R*	Ref.	N—C_1 (N—C_4)	C_4—C_5 (C_5—C_1)	C_1—C_2 (C_3—C_4)	C_2—C_3	C_5—*R*
11	Bis(imidazole)-tetraphenylporphine	$(FeCl)^{++}$	C_6H_5	[825]	1.382 (1.375)	1.392 (1.391)	1.434 (1.439)	1.350	1.498
12	Tetraphenyl-porphine	$(FeO)^{++}$	C_6H_5	[826]	1.376 (1.378)	1.408 (1.393)	1.452 (1.439)	1.345	1.490
13	Tetraphenyl-porphine	$FeOH^{++}$ (H_2O)	C_6H_5	[818]	1.356 (1.383)	1.386 (1.442)	1.409 (1.442)	1.352	1.494
14	Tetra(4-pyridyl-) porphine	Zn^{++}	C_5H_4N	[827]	1.369	1.406	1.447	1.355	1.491
	Averages Nos. 2—14				1.375	1.397	1.438	1.356	1.499
15	Etioporphyrin	Ni^{++}	H	[828]	1.396	1.398	1.427	1.335	—
16	Diacetyldeutero-porphyrin-IX-dimethylester	Ni^{++}	H	[829]	1.383	1.375	1.446	1.350	—
17	Oktaethyl-porphyrin	Ni^{++}	H	[830]	1.396 (1.397)	1.373 (1.372)	1.453 (1.444)	1.362	—
18	Mesoporphyrin-IX-dimethylester	$(FeOCH)_3^{++}$	H	[831]	1.383 (1.408)	1.392 (1.363)	1.468 (1.464)	1.368	
19	Deoxophyllo-erythroetio-porphyrin-1,2-dichloro-ethane	VO^{++}	H	[832]	1.38 (1.39)	1.38 (1.37)	1.44 (1.45)	1.34	
	Averages Nos. 15—19				1.392	1.378	1.449	1.351	

The polymethine structure is prominent in the *X*-ray structural data. To demonstrate this, the average values of all equivalent bond lengths of the tetraphenyl porphines have been presented in (6.52). Similar interatomic distances are likewise realised in other porphine derivatives as shown in Table 14 under Nos. 15 to 19. All C—N bond lengths are remarkably equal which does not admit of a localisation of the two protons or the

Average bond lengths taken from Table 14 (6.52)

metal ions on specific nitrogen atoms. The bond lengths in the trimethinecyanine fragments are typically polymethine-like, with C—C bonds of about 1.39 Å and the C—N distances extended which, as in other cyanines (Section 3.1), is attributable to the branching effect over the pyrrole rings. The branching causes a polyene-like character of the remaining C—C bonds of the pyrrole rings with bond lengths modified to 1.44 Å in the direction of a single bond (1.47 Å for C_{sp^2}—C_{sp^2}), and to 1.36 Å, in the double-bond direction (1.33 Å, cf. Table 1). The planes of the phenyl substituents are twisted through an angle of almost 90° relative to the plane of the porphine ring, thus preventing resonance between the different ring systems. Hence the C_5—R bond lengths are about 1.50 Å. An exception to this is represented to a certain degree by the ferric hydroxide chelated compound (Table 14 No. 13), in which a certain polyene-like structure is formed in the polymethine ring.

The bond type typical for cyclically coupled polymethines disappears in the tetroxoporphinogene (6.53), in which the aromatic character of the pyrrole ring dominates with

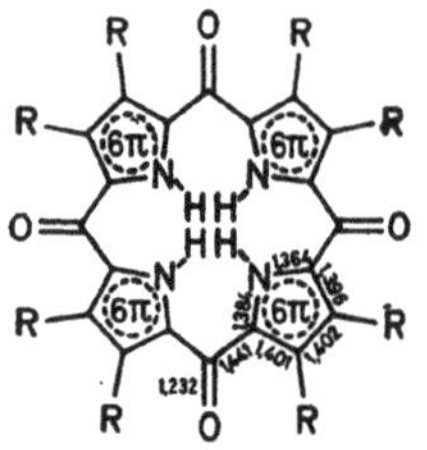

(6.53) [833]

bond lengths of about 1.40 Å. The slightly enlarged C=O bond length (1.23 Å) and the reduced C—C bond length (1.44 Å) within the porphine ring indicate, however, that the macrocyclic system is slightly involved in the resonance yet.

A porphine-like chromophoric system without pyrrole ring branching is contained in the tetrabenzo[b, f, j, n]-[1, 5, 9, 13]-tetraazacyclohexadecine (6.54), of which the nickel iodide and boron fluoride complexes have been subjected to *X*-ray analyses. Formula (6.54) shows the average values of identical bonds of the boron fluoride complex.

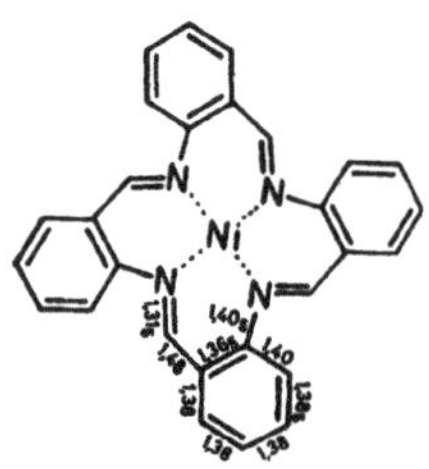

(6.54) [834]

The condensed benzene rings induce in this case an unsymmetrical meropolymethine-like electron distribution in the trimethinecyanine fragments. This gives rise to a double-bond length of 1.36 Å in the benzene rings, which may counterbalance the tendency of aromatisation. The tribenzo[b, f, j]-[1, 5, 9]-triazacycloduodecine-nickel-II-nitrate [835] is built on a similar structural principle, but the structural data are not accurate enough to allow a definite statement to be made to this effect.

Similar asymmetric structures are present in the phyllochlorine ester [836], with one pyrrole ring being partially hydrogenated. This again leads to alternating, mero-polymethine-like bond lengths. An even more pronounced asymmetry is to be found in the nickel-(II)-1,8,8,13,13-pentamethyl-5-cyano-trans-corrin-chloride (6.55). Here, all

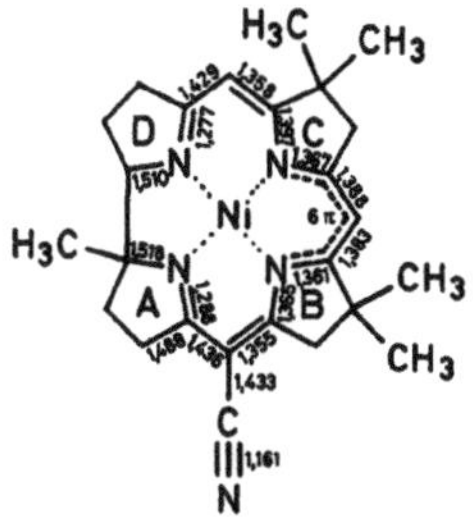

(6.55) [837]

the pyrrole rings are hydrogenated, thus cancelling the branching effects of the polymethine system over the polyene chains. Consequently the bond lengths of the pyrrole rings, not indicated in (6.55), range between 1.48 and 1.57 Å. Only one highly symmetrical trimethinecyanine structural group can be identified in the molecule. This apparently interacts with the two meropolymethine-like structural groups, so that the average C—N bond length of the trimethinecyanine group remains comparatively large at 1.36 Å. The C=N bond lengths in the two meropolymethine groups are strongly reduced to 1.28 Å, which is consistent with the polyene-like electron distribution.

The nickel complex with the *A*/*D*-seco-corrinoid ligand [838] has practically the same structure, in which the cyclic system between the *A* and the *D* rings is opened. By contrast, in the analogous platinum and palladium complexes [838] an unsymmetrical polyenic structure is induced in the remaining trimethinecyanine system as well, as shown by the palladium compound in (6.56). Even more significant is the elimination

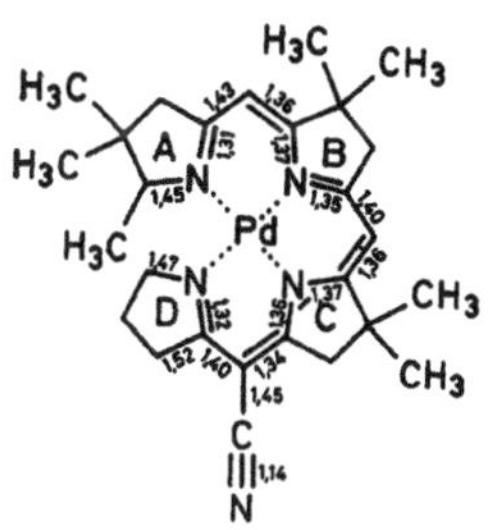

(6.56) [838]

of the polymethine structures in the *C*/*D*-seco-corrinoid-nickel-II-complex [839] (opening of the macrocyclic ring of (6.55) between rings *D* and *C*), in which it is possible to formulate only two unsymmetrically interacting trimethinemerocyanine structural groups.

The phthalocyanines (6.57) may be considered as cyclically coupled aza-polymethine-cyanines. Compared to the linear azatrimethines ((3.21) and (3.22), Section 3.1), a significant enlargement owing to the branching effect is to be observed in the C=N bonds of the pyrrole rings. The polyene-like double bonds in the pyrrole rings are

Table 15: Average Bond Lengths [Å] In Phthalocyanines (6.57)

No.	Substance	*Me*	Ref.	$N-C_1$ ($N-C_4$)	C_4-N_5 (C_1-N_5')	C_1-C_2 (C_3-C_4)	C_2-C_3	C_3-C_6 (C_2-C_9)	C_6-C_7 (C_8-C_9)	C_7-C_8
1	Phthalocyaninato-tin(II)	Sn^{++}	[840]	1.373	1.328	1.456	1.391	1.395	1.386	1.400
2	Dichloro-phthalocyaninato-tin(IV)	$(SnCl_2)^{++}$	[841]	1.378	1.333	1.454	1.402	1.386	1.386	1.385
3	Phthalocyaninato-lead(II)	Pb^{++}	[842]	1.375	1.35	1.49	1.45	1.39	1.44	1.38
4	Phthalocyaninato-copper(II)	Cu^{++}	[843]	1.366	1.330	1.453	1.400	1.390	1.377	1.413
5	Phthalocyaninato-pyridine-manganese(II)	Mn^{++}	[844]	1.37	1.33	1.45	—[a])	—	—	—
6	Bis-phthalocyaninato-uranium(IV)	U^{4+}	[845]	1.38	1.32	1.46	1.40	1.40	1.39	1.41
	Averages			1.374	1.331	1.460	1.409	1.392	1.396	1.398

[a]) bond lengths not indicated

weakened by comparison to the porphines (6.52) in consequence of the aromatisation tendency of the condensed benzene rings. Some more recent *X*-ray data are summarised in Table 15; the calculated average bond lengths are given in (6.57).

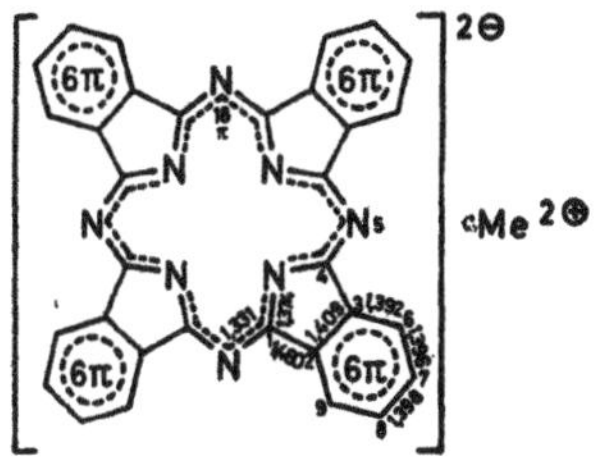

Average bond lengths taken from Table 15 (6.57)

7. The Structure of Polymethine Radicals

7.1. Radical Cations

Organic compounds which contain a structural group as in (2.1) with an even number of chain atoms are likewise of a deep colour. They possess an odd number of π-electrons in consequence of the $(N \pm 1)$ π-electron occupation. Hence we are concerned with organic radicals. These are termed polymethine radicals as proposed by DÄHNE [74]. According to the quantum-chemical *HMO* model [75—77], an ideal polymethine radical with carbon atoms in the chain and nitrogen terminal atoms and an $(N + 1)$-π-electron occupation (2.1, $Y = C$, $X, X' = NR_2^{1/2\oplus}$, $N = 2, 4, 6 \ldots$) ought to exhibit equal bond orders p_{rs} between centres r and s of neighbouring atoms and alternating π-electron densities q_r, in analogy to the odd-numbered polymethine dyes. Only at the central bond of the radical chains is the π-electron alternation perturbed. The values can be calculated as a function of the number of atomic centres N according to the following formulae given by FABIAN et al. [76]

$$p_{rs} = \frac{1}{N} \operatorname{ctg} \frac{\pi}{2N} \qquad q_r = 1 - \frac{(-1)^r}{N} \cdot \operatorname{ctg} \frac{\pi}{2N} (2r - 1)$$

with $r = 1, 2 \ldots N$

Although a variety of deeply coloured polymethine radicals is known [74], no syntheses have been developed to date for their simplest representatives, in conformity with formula (2.1, N = even number) with unbranched chains. The most widely known polymethine radicals with a double branched polymethine chain are WURSTER's coloured salts (7.2) which can be obtained by oxidation of the para-phenylenediamine (7.1) or by reduction of the quinonediimmoniumdication (7.3). Presented in (7.2) are the average values of the bond lengths determined by three accurate X-ray analyses of WURSTER's Blue ($R = R' = CH_3$, $X^\ominus = J^\ominus$, $ClO_4^\ominus$). Contrary to the expected

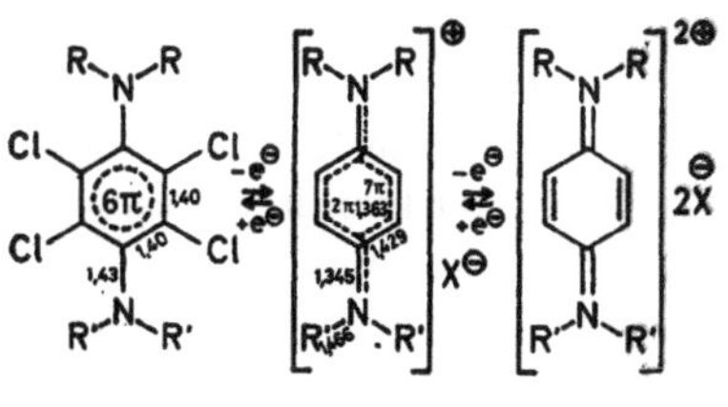

(7.1) [846] (7.2) (7.3)

Averages of (7.2) were taken from [847—849]

values of the *HMO* model [76, 77] for linear polymethine radicals, the bond lengths are not ideally equalised.

Apparently owing to the branching effect a more polyene-like electron distribution results in the ring. At 1.36 Å the central C=C bonds are shortened while the C=C bonds at the branching points are extended to 1.43 Å, without however reaching the values of a true polyene. For comparison, we refer to the structure of the para-benzoquinone (6.6, Section 6.1) with bond lengths of 1.47 and 1.32 Å in the ring and to the value of a C=N double bond, 1.30 Å (Table 1), as no *X*-ray structure analysis is available for (7.3). WURSTER's Red (7.2, R = H, R' = CH_3) [850] has in principle the same structure. The fact that the polymethine radicals have a unique structure is supported by a comparison to the para-phenylenediamine (7.1, $R = R'$ = H; 2,3,5,6-tetrachloro substituted) or its hydrochloride [851] which, as expected, must be regarded as substituted aromatic compounds on account of the relatively large C—N distances and the approximately identical ring bonds. This is equally true of the benzidines [852, 853] which are of interest as reference substances in the following discussion of vinylene homologous polymethine radicals.

Polymethine radical-like structures of the type of WURSTER's Blue are to be found in the radical ion (7.4) of the 4,4'-bis(dimethylamino)biphenylamine chlorate and perchlorate [855], respectively, and as mentioned in Section 3.1, in a certain sense

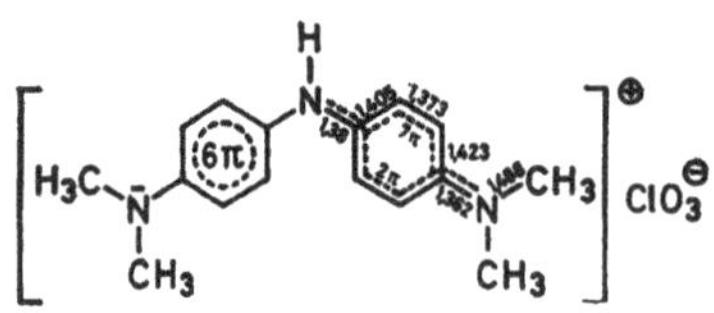

(7.4) [854]

in methylene blue (3.24). In both types of compound a tetramethine-radical structural group corresponding to the WURSTER's radicals can be formulated over one half of the molecule in each case or an aza-nonamethinecyanine fragment over the total chromophore. A comparison of the bond lengths to those of WURSTER's Blue shows, however, that in (7.4) the branched tetramethine-radical structure is mainly preferred.

Hexamethine and tetramethine radical structures would be expected to occur in the N,N-dimethyl-4,4'-dipyridinium cation (7.5) and in the corresponding 2,2'-dipyridinium derivatives. As in the WURSTER's dyes, the radical structure should be intermediate between that of the dipyridils [856] and that of the dipyridinium dications [857—865].

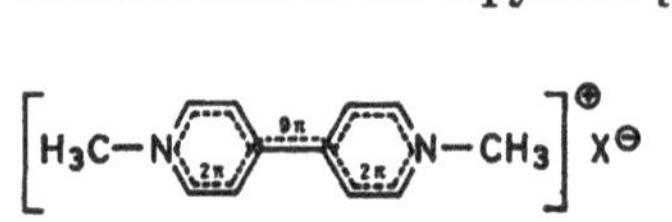

(7.5)

A discussion of the bonding and structure is complicated by the inaccuracy of the *X*-ray data available and by the lack of reliable reference data for the dipyridils. Yet the central C—C bond length in some metal complexes which seem to have an intermediate polymethinic structure [866—868] is found to be shorter than in the aromatic benzidines [852, 853].

7.2. Radical Anions

A situation analogous to that in the WURSTER's radicals exists in the polymethine radical-anions of the type tetracyanoquinodimethane (7.7). Here, too, there are redox equilibria both to the polyene structure, the tetracyanoquinodimethane (*TCNQ*) (7.6) and to the aromatic structure, the dianion of the tetracyanoquinodimethane (7.8). Structure analyses are available of (7.6) and (7.7) [859, 870—874], of which Table 16

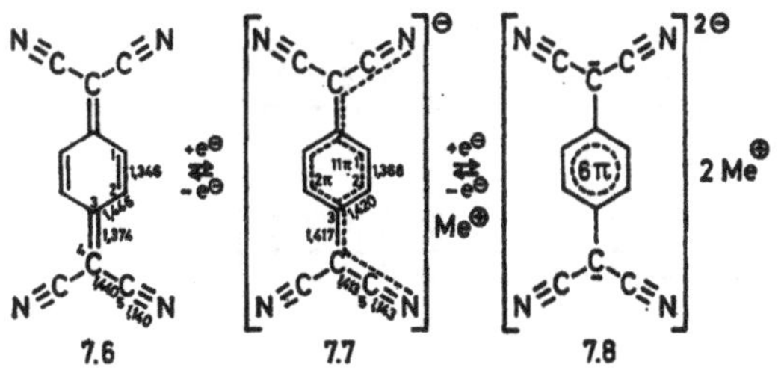

(7.6) [869] (7.7) (7.8)

Average values of (7.7) were taken from Table 16 Nos. 2 to 5

contains some values of interest. The structure of tetracyano-naphthoquinodimethane has also been determined [883]. Similar to the WURSTER's radicals, the tetracyanoquinodimethane radical (7.7) displays a clearly diminished polyene structure, without attaining an ideal polymethine structure. An exception to this rule being the C≡N triple bonds — as in the cyanopolymethines (Section 3.1) — whose bond lengths scarcely depend on the specific electronic state.

Partially hydrogenated TCNQ compounds of the type $(TCNQ)_2^{\ominus}$ [884—888]; $(TCNQ)_3^{2\ominus}$ [875, 876, 877, 889] and others [858] apparently take an intermediate position between the quinone structure (7.6) and the polymethine radical structure. Here two kinds of molecule appear in the crystals, of which, to the surprise of the authors [875], only one kind of molecule carries the negative charge in most of the cases. An explanation is provided by the polymethine concept: as is to be expected, the uncharged molecule has a more polyene-like structure (7.6), while the charged molecule is closer to the polymethine-like structure (7.7). The averaged bond lengths are summarised for these compounds in Table 16, Nos. 6, 7 and 8.

7.3. Complexes With Polymethine Radical-like Structures

It is well-known that the electronic structure of a charge-transfer complex is deformed in the ground state in the direction of a transfer of electrons from the donor to the acceptor [890]. This deformation facilitates the transfer of electrons between the partners during light absorption, thus accounting for the colour of the complexes. Opinions diverge on the factors that effect the arrangement and the structure of the complex partners in the crystal lattice [891a, b]. According to the polymethine concept, a partial transfer of an electron between the complex partners ought to induce in the donor molecule a structure analogous to that of the polymethine radical-cations, while in the acceptor molecule a structure comparable to that of the polymethine

Table 16: Bond Lengths [Å] In Tetracyanoquinodimethane (7.6) and Its Polymethine Radical (7.7)

No.	Formula		Charge	Cation	Ref.	$C_1{=}C_2$	$C_2{-}C_3$	$C_3{=}C_4$	$C_4{-}C_5$	$C_5{\equiv}N$
1	TCNQ		± 0	—	[869]	1.346	1.448	1.374	1.440	1.140
2	$(TCNQ)^{\ominus}$		$\ominus$	$Na^{\oplus}$	[871]	1.354	1.427	1.420	1.420	1.142
3	$(TCNQ)^{\ominus}$		$\ominus$	$K^{\oplus}$	[872]	1.37	1.42	1.40	1.41	1.13
4	$(TCNQ)^{\ominus}$		$\ominus$	$Rb^{\oplus}$	[873]	1.374	1.423	1.420	1.416	1.153
5	$(TCNQ)^{\ominus}$		$\ominus$	$H_2\overset{\oplus}{N}(CH_2{-}CH_2)_2O$	[874]	1.376	1.411	1.428	1.405	1.145
	Averages $(TCNQ)^{\ominus}$					1.368	1.420	1.417	1.413	1.143
6	$(TCNQ)_2^{\ominus}$	Form a	± 0	—	[875]	1.341	1.444	1.371	1.428	1.140
		Form b	$\ominus$	$NH(C_2H_5)_3^{\oplus}$		1.355	1.427	1.410	1.419	1.152
7	$(TCNQ)_3^{2\ominus}$	Form a	± 0	—	[876]	1.345	1.444	1.388	1.422	1.153
		Form b	$\ominus$	$Cs^{\oplus}$		1.376	1.433	1.409	1.408	1.158
8	$(TCNQ)_3^{2\ominus}$	Form a	± 0	—	[877]	1.338	1.440	1.366	1.427	1.138
		Form b	$\ominus$	$NH(CH_3)_3^{\oplus}$		1.365	1.410	1.419	1.414	1.146
	Averages	Form a				1.341	1.443	1.375	1.426	1.144
	Averages	Form b				1.365	1.423	1.413	1.414	1.152
9	TCNQ + *TTF*[b])		1:1 Complex		[878]	1.356	1.433	1.402	1.423	1.151
10	TCNQ + *TMPD*[a])		1:1 Complex		[879]	1.373	1.414	1.406	1.412	1.160
11	2 TCNQ + *TMPD*[a])		2:1 Complex		[880]	1.355	1.434	1.395	1.426	1.171
12	TCNQ + Durol		1:1 Complex		[881]	1.358	1.462	1.391	1.445	1.152
13	TCNQ + Anthrazene		1:1 Complex		[882]	1.355	1.462	1.370	1.428	1.132
	Averages of Complexes 9—13					1.359	1.441	1.393	1.427	1.153

a) *TMPD* = N,N,N′,N′-Tetramethyl-para-phenylene diamine
b) *TTF* = 2,2′-Bis-1,3-dithiole

radical-anions ought to be induced. This is actually the case. In the 1:1 complexes of TCNQ (7.6) with N,N'-tetramethyl-paraphenylenediamine (7.1, $R = R' = CH_3$) [879], 2,2'-bis-1,3-dithiole [878], N-methylphenazine [892], or with N,N'-dimethylbenzimidazole [893], the radical-ion structures are nearly ideally realised in both complex partners (cf. Table 16, Nos. 9, 10) which points to a nearly complete transfer of an electron from the diamine to the TCNQ. For this reason, these and similar compounds have become very important as organic conductors with quasi-metallic behaviour [884, 894]. Also in the 2:1 complex of the same molecules depicted in (7.9) the two complex

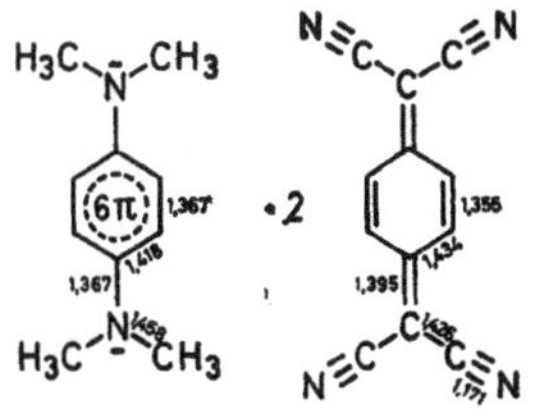

(7.9) [880]

partners hold a position intermediate between the polymethine radical structure and the aromatic structure (7.1) for the TMPD and between the radical structure and the polyene structure (7.6) for the TCNQ, respectively, with the extension of the C≡N bond to 1.17 Å being particularly striking in the TCNQ. As expected, also the exo double bonds of the ring are somewhat lengthened in comparison to TCNQ (7.6). In the para-phenylenediamine part on the other hand the C—N bond length of the dimethylamino group to the ring of 1.37 Å expectedly lies between that of the Wurster's Blue perchlorate (7.2) and that of the non-complexed para-phenylenediamine (7.1).

The result is similar in the 1:1 complexes of N,N'-tetramethyl-para-phenylenediamine and chloranil or tetracyanobenzene [895—897], of which the former is presented in (7.10).

The Wurster's Blue structure again can be readily identified in the para-phenylenediamine part of the complex. In the quinone part, too, there are indications of an

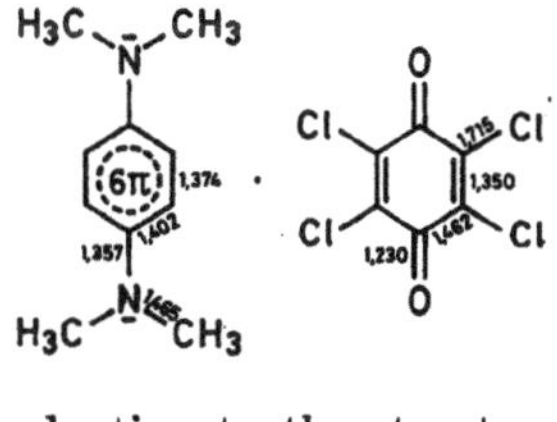

(7.10) [897]

adaption to the structure of the semiquinone radical-anion with a certain degree of bond length equalisation, as a comparison with the chloranil structure (7.11) will show. Analogously, a partial benzidine radical-cation structure is induced in the chloranil complexes of the N,N'-tetramethylbenzidine, as demonstrated by the structures of

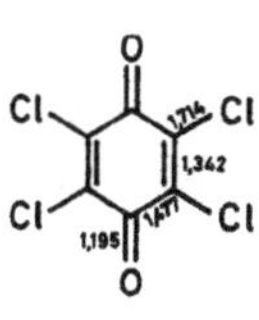

(7.11) [715]

the 1:2 [898] and 1:1 complexes [899]. The central C—C bond and the C—N bonds at 1.44 and 1.37 Å, respectively, are appreciably shorter than in the complex-free benzidine, in which the adequate, mean bond lengths are 1.51 and 1.42 Å, respectively [852, 853]. Comparing the 2:1 to the 1:1 complex, there occurs an increase in bond length equalisation and hence in the latter an increase in the polymethine radical character, as expected. Similar results were attained with other complexing reagents, such as TCNQ or s-trinitrobenzene [900—902].

The structures of many other complexes of tetracyanoquinodimethane have been determined [878—882, 903—905], of which only a few examples are given in Table 16. The average values obtained from five *X*-ray analyses clearly show the central double bond of the quinone ring to be enlarged compared to the complex-free TCNQ. Also significant seems to be the lengthening of the C≡N triple bond. The alterations are smallest — and in part reversed — in the $C_2—C_3$ single bonds at the branching points in the ring and in the adjoining exo-$C_3—C_4$ double bonds which again may be associated with the peculiarities of the branching effect. But as a rule it is confirmed that the structural changes of the complex partners are in the direction of a more polymethine-like structure.

Structures analogous to the polymethine radicals may be assumed to exist in the deeply coloured metal complexes of the dithiolenes (7.12) [906—908], of glyoxime

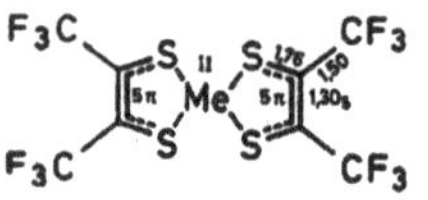

(7.12) [908]

[909—913] and similar compounds [914, 915] as well as in the 1,10-phenanthrolines (7.13) [916—918]. In analogy to the chelates of the acetylacetone of a trimethine structure (Section 3.2) dimethine structural groups with sulphur and nitrogen terminal atoms, respectively, are realised in these complexes. True radicals are not formed, owing to an intramolecular compensation of the electron spins.

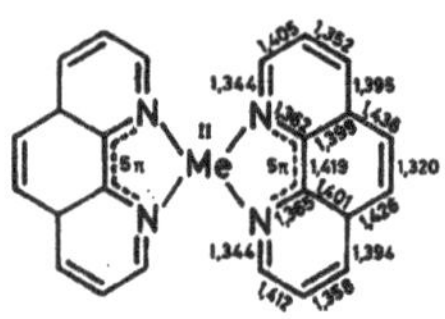

(7.13) [916]

Here again no reliable structural data of the non-complexed compounds are available for reference, so that a thorough structure-chemical discussion is not possible. Noticeable in the phenanthroline complexes is the fixation of the double bonds depicted in formula (7.13) which one would not have expected in this form. Presumably branching effects are again involved here.

7.4. Other Radicals

The violet 2,4,6-triphenylverdazyl radical (7.14) represents a non-ionic radical with formally N π-electrons on N chain-atoms (with $N = 3$). The pronounced equilisation of bond lengths in the verdazyl ring indicates delocalisation to a large extent of the

π-electrons involving the additional nitrogen atoms and the methylene group that has a C—N bond length of 1.44 Å which is significantly shorter than that of a true single bond of 1.47 Å (cf. Table 1). In consequence the ring seems to have a polymethine-like structure of 7 π-electrons on 6 chain-atoms which accounts for the deep colour of the

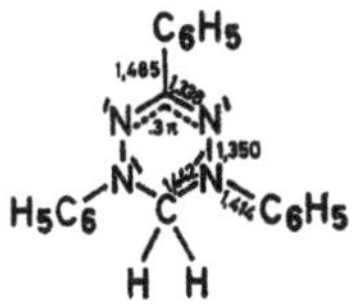

(7.14) [919]

compound. It remains unclear whether the required extra π-electrons are provided by the n-electrons of the nitrogen atoms or as a result of hyperconjugation of the methylene group.

8. Nonalternant Hydrocarbons of Polymethinic Structure

As demonstrated, almost all coloured, unsaturated organic compounds contain more or less pronounced polymethine structural groups of the general formula (2.1). An exception of principle is represented by the so-called nonalternant hydrocarbons[1] which like aromatic compounds contain N π-electrons on N atomic centres, yet absorb at appreciably longer wavelengths. The best known example of this is azulene (8.1) which having only 10 π-electrons is of blue colour and absorbs even in the near infrared

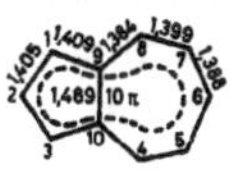

Average bond lengths taken from Table 17 (8.1)

spectral region. As early as in the formulation of the concept of the polymethine state [6] it was pointed out that there is evidence of analogies between the electronic structure of the nonalternant hydrocarbons and that of the polymethines. While in azulene the bond length equalisation may also be interpreted in terms of an aromatic structure, the strongly alternating charge density distribution over the molecule, which is reversed on light absorption [920], is indicative of a polymethine-like behaviour. In both the polymethines and in the nonalternant hydrocarbons [8] charge density alternation causes a preferred ability of electrophilic substitution at atoms of high π-electron density and a corresponding ability of nucleophilic substitution at atoms of low π-electron density. Investigations of the *IR*-spectra by Mitzinger [98] and Seydewitz [99] have proved the C—H vibrational frequencies and the respective intensities depend in both types of compound on the charge density of the methine atoms in question. The dependency of the ^{13}C—H coupling constants on the electron density on the ^{13}C atom detected by Radeglia et al. [96, 97] in the polymethines is also to be found in azulene [98]. It has been suggested by the author [98] that it should be possible to detect angle deviations in the azulene, same as have been found in the polymethine chains [64], due to rehybridisation effects which are determined by the π-electron density.

[1] In the literature nonalternant hydrocarbons are defined as unsaturated carbocycles, which do not allow subdividing the carbon atoms possessing π-electrons into two groups such that each atom of the one group has only neighbouring atoms of the other. The term "nonalternant hydrocarbons" is not very well chosen in view of the strongly alternating π-electron density distribution in these compounds.

These analogies support the attempt to discuss from a general point of view the colour of unsaturated organic compounds based on the charge density alternation typical of the polymethine state. Apparently, the specific atomic arrangement in the nonalternant hydrocarbons has the same effect as has, in polymethines, the over- or underpopulation by π-electrons of a linear chain of suitably hybridised atoms.

However, as already mentioned in the introductory Chapter 2., it is not possible to distinguish by X-ray analysis between aromatic and polymethinic bond lengths. Yet, for the sake of completeness we decided to list in the following a few structural data of nonalternant hydrocarbons.

The bond lengths obtained from various structure analyses of azulene and its derivatives are summarised in Table 17. Azulene itself is disordered in crystals [921, 922]. More accurate data can be taken from the structure analysis of the azulene-1,3-dipropionic acid [924], the distyrylazulene [925], or from the electron diffraction analysis [923]. The structure analyses confirm that the bond lengths are equalised along the peripheral polymethine chain, whereas the central bond common to both rings is significantly extended with an average value of 1.49 Å compared to the adequate bond length in the aromatic naphthalene (1.39 Å).

There is no substantial change in the interatomic spacings on complexing the azulene with s-trinitrobenzene (Table 17, No. 6). Therefore the average values of the first six structures in the Table have been presented in formula (8.1). By comparison, the molecular structure of the azulene is strongly modified when complexing with the metal atoms of the subgroup elements. Along with this, the colour of azulene diminishes. In the compounds 7 to 13 of Table 17, the five-membered ring is ferrocene-like co-ordinated with the metal atom, thereby shortening significantly in particular the central C_9-C_{10} bond. In the seven-carbon ring there occurs either an additional co-ordination with a second metal atom, which effects an extension of the C_4-C_{10} and C_8-C_9 bonds while maintaining the bond length equalisation over the atoms C_4 to C_8 in the sense of formula (8.2) (Table 17 Nos. 7, 8).

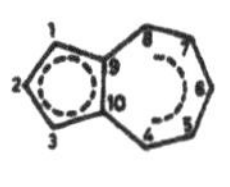

8.2

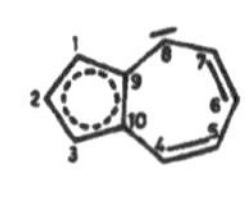

8.3

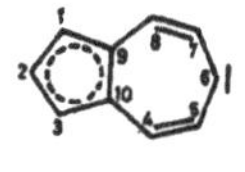

8.4

Or else polyene-like structures in terms of formulae (8.3) or (8.4) (Table 17 Nos. 9—11 and 12, 13, respectively) are induced. In the ruthenium complex (Table 17 Nos. 14, 15), finally, the ferrocene complexing at the five-membered ring is eliminated. The central C_9-C_{10} bond regains its original length and, depending on the crystal modification, double bonds are formed in either the 4—5 and 7—8 position (Table 17 No. 14) fitting formula (8.4) or in the 2,3-position only (Table 17 No. 15).

Little is known about the structures of more complex nonalternant hydrocarbons. The structure of the 6-dimethylamino-5-aza-azulene [934] shows the expected deviations from the structure of azulene itself in accordance with formula (8.4). In the 5,6-dimethyl-2-phenylcyclopent-[c, d]-azulene (8.5) the azulene structure is evident, with an ethylene structural group that links the two rings together. This is however

Table 17: Bond Lengths [Å] In Azulene (8.1) and Its Derivatives

No.	Compound	Ref.	C_1-C_2 (C_2-C_3)	C_1-C_9 (C_3-C_{10})	C_8-C_9 (C_4-C_{10})	C_4-C_5 (C_7-C_8)	C_5-C_6 (C_6-C_7)	C_9-C_{10}
1	Azulene	[921]	1.383	1.389	1.373	1.405	1.403	1.481
			(1.488)	(1.425)	(1.387)	(1.431)	(1.276)*	
2	Azulene	[922]	1.363	1.378	1.395	1.440	1.422	1.483
			(1.425)	(1.459)	(1.369)	(1.337)	(1.337)	
3	Azulene (by electron diffraction)	[923]	1.399	1.418	1.383	1.406	1.403	1.501
4	Azulene-1,3-dipropionic acid	[924]	1.395	1.413	1.373	1.402	1.382	1.490
			(1.412)	(1.394)	(1.386)	(1.384)	(1.387)	
5	Cis,cis-1,3-Bis(styryl)azulene	[925]	1.400	1.416	1.389	1.391	1.380	1.479
6	Azulene,s-Trinitrobenzene Complex	[926]	1.398	1.404	1.390	1.393	1.391	1.498
			(1.387)	(1.395)	(1.392)	(1.402)	(1.397)	
	Averages		1.405	1.409	1.384	1.399	1.388	1.489
7	Guaiazulene-dimolybdenum-hexacarbonyl	[927]	1.423	1.424	1.430	1.410	1.405	1.434
			(1.384)	(1.413)	(1.485)	(1.394)	(1.405)	
8	trans-Azulene-dimanganese-hexacarbonyl	[928]	1.445	1.444	1.476	1.425	1.427	1.419
			(1.417)	(1.437)	(1.476)	(1.396)	(1.418)	
9	Diazulene-tetrairon-decacarbonyl	[929]	1.424	1.424	1.435	1.364	1.434	1.424
10	dimeric Azulene-tricarbonylmethyl-molybdenum	[930]	1.385	1.412	1.462	1.509	1.344	1.439
			(1.410)	(1.415)	(1.510)	(1.353)	(1.441)	
11	Bis-π-azulene-iron Molecular form a	[931]	1.406	1.432)	1.436	1.500	1.350	1.441
			(1.397)	(1.436)	(1.524)	(1.324)	(1.447)	
12	Molecular form b		1.419	1.452	1.456	1.296	1.503	1.445
			(1.428)	(1.447)	(1.465)	(1.360)	(1.474)	
13	Azulene-diiron-pentacarbonyl	[932)	1.415	1.422	1.441	1.325	1.443	1.452
			(1.418)	(1.430)	(1.453)	(1.391)	(1.432)	
14	4,6,8-Trimethylazulene-tetraruthenium-ennea-carbonyl a monoclinic form	[933]	1.452	1.401	1.458	1.307	1.434	1.539
			(1.466)	(1.399)	(1.610)	(1.311)	(1.522)	
15	b triclinic form		1.424	1.425	1.462	1.416	1.486	1.472
			(1.368)	(1.449)	(1.472)	(1.415)	(1.462)	

* Obviously a misprint in the Literatur [406]. This should probably read 1.376 Å.

involved in the resonance of the chromophoric system, as is proved by the partial extension of the double bond from 1.33 Å to 1.36 Å and the partial contraction of the single bonds from 1.47 Å (cf. Table 1) to 1.42 Å. As a result, the central C_9-C_{10} bond of

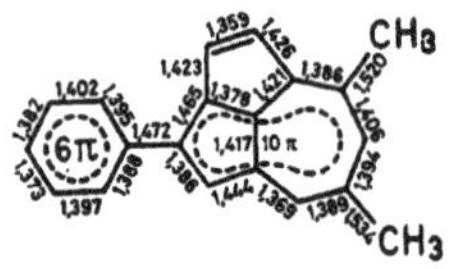

(8.5) [935]

the azulene is apparently reduced from 1.48 Å to 1.42 Å. Not affected, on the other hand, is the aromatic structure of the phenyl substituent.

Surprisingly, there is a one-sided preference for an azulene structure in the black crystals of the 4-methylpentaleno[6, 6a, 1, 2-def]-heptalene (8.6), so that a more polyene-like structure is induced in the remaining molecule. Quantum-chemical SCF calculations

(8.6) [936]

predicted a uniform alternation of all bonds [936]. Apparently the system avoids symmetrical branching, favoured by steric strains, in the sense of the rules outlined at the beginning.

9. Summary and Outlook

The bond lengths in unsaturated organic compounds determined by *X*-ray analyses provide further evidence of the existence of an ideal polymethine state characterised by unique features verify the applicability of the concept of intermediates existing between the three ideal limiting states, the polymethine state, the polyene state and the aromatic state. Over and beyond this, *X*-ray analysis has brought forth new, unexpected results, which give a deeper insight into the molecular and electronic structures of coloured organic compounds.

Few rules are required for an understanding of the main structural principles of these compounds, provided that they are sterically not hindered. As far as possible, there will be realised in a compound as many as possible independent, energetically stabilised, ideal aromatic structural groups, especially π-electron sextets, along with as many as possible independent and energetically stabilised, ideal polymethine structural groups, both of them with interatomic distances of about one-and-a-half bond. If these ideal structures are perturbed, say by branching of the conjugated system or by linkage among differing structural groups or by growing asymmetry of the ideal fragments due to substituent effects, the real structure then seems to be normally determined by PAULING's rule that the sum of bonds originating from an atom is constant. This explains the occurrence of polyenic bonds, that is bonds having a greater single- or double-bond character, in substances with a more extensive conjugated π-electron system. Deviations from PAULING's rule may occur in coupled polymethines, where bond lengths are additionally changed by Coulomb forces in consequence of perturbation of the charge density alternation typical of ideal polymethines.

Proof of the alternating π-electron density distribution has been established indirectly by alternations in the bond angles of sp^2-hybridised atoms and by peculiarities relating to the intermolecular interactions of polymethines within the crystal lattice. Another way of determining electron density alternation in the future is the determination of as accurate C—H distances as possible.

In spite of the large amount of structural data available there remain a great many central key compounds, of which no *X*-ray analyses have been made. The structures of the following types of compound should receive particular attention in future studies:

1. long-chain, ideal polymethines
2. singly branched, ideal polymethines
3. singly unbranched and branched meropolymethines

4. structurally unsymmetrical polymethines with symmetrical π-electron distribution
5. simple model compounds of ideal polymethine-aromatic intermediates
6. ideally coupled polymethinecyanines
7. increasing the accuracy of structural data on indigoid dyes
8. simplest, still to be synthesised, unbranched polymethine radicals

For further extending structural chemistry, crystal structure analyses of the greatest possible accuracy are of great relevance. This requires first of all that

1. the accuracy of experimental data be increased
2. the location of the hydrogen atoms be determined with higher precision
3. that further attempts be made to determine the electron density distribution at the atoms

In developing the methodology of X-ray analysis efforts are being made primarily towards perfecting and automating experimental equipment, towards more sophisticated and improved mathematical and physical methods as well as the use of computers and programming facilities. Neutron diffraction methods, too, hold great promises for obtaining even more accurate structural data.

10. Addenda

The results of a few publications which have appeared in the meantime should be stressed in addition.

1. Stimulated by the latest work of Fabian et al. on coupled polymethines [705] we have given a more precise definition on branched polymethines and on coupled polymethines taking into account the results of our work presented herein [695b].

Branched polymethines as well as *non-branched, ideal polymethines* are termed *alternant polymethines* as they are characterised by a regularly alternating π-electron density distribution. Branched polymethines as a rule absorb at shorter wavelengths than their parent non-branched systems.

Coupled polymethines as well as *ideal polymethine radicals* are termed *nonalternant polymethines*. The π-electron density alternation which is characteristic for ideal polymethines is perturbed in these systems. Depending on the coupling conditions one can distinguish between different types of nonalternant polymethines; but in all types the light absorption of the longest wavelength transition is pronouncedly red-shifted in comparison to the longest wavelength absorption band of the constituent single polymethine fragments.

2. Allmann et al. have determined the X-ray structures of the 1,1′,3,3,3′,3′-hexamethylindocyanine tetrafluoroborate (10.1) and of the meso-cyclopropyl-(1,1′,3,3,3′,3′-hexamethyl)-indodicarbocyanine tetrafluoroborate (10.2). The structural data agree well with those of other cyanines cited in Section 3.1.

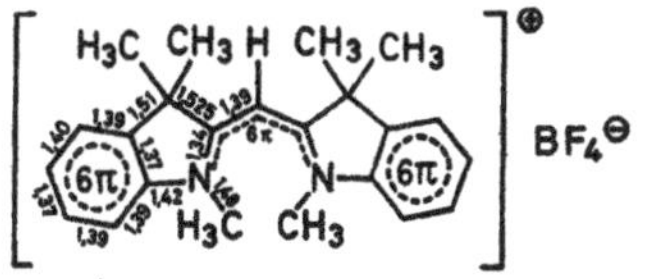

(10.1) [937]

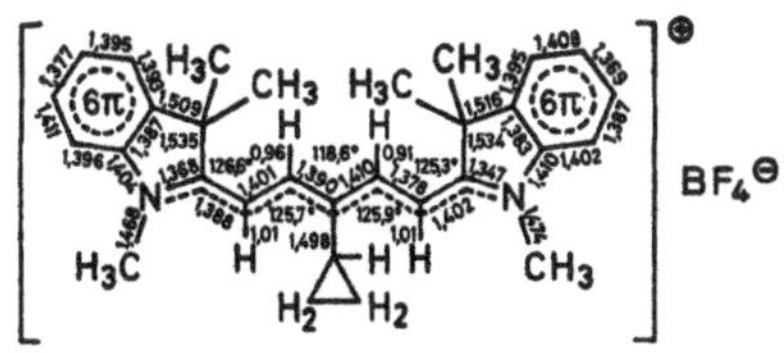

(10.2) [938]

A comparison of the data of (10.1) and (10.2) reveals the relationship between steric hindrance and polymethinic angle alternation discovered by KULPE et al. [64] in more complicated cyanines: only the angles at the methine atoms in the α-position to the heterocycles, which should be smaller than 120° owing to the high π-electron density, are widened up to 126° in consequence of the steric hindrance by the hydrogen atoms of the methine atoms in the β-position. On the other hand the angle at the central, negatively charged methine atom of (10.2) which shows no steric hindrance has the expected low value of 118.6° [64].

Also the partial fixation of a non-symmetrical valence bond structure in consequence of polarisation forces in the crystal lattice found by us in simple cyanines (cf. Section 3.1) is confirmed by the structure (10.2). The unsymmetrically situated anions in the crystal lattice induce a week alternation of the bond lengths in the polymethinic chain of the dye molecule.

Finally there seem to be significant differences in the C—H distances of the methine groups depending on their electronic charge. As to be exptected theoretically (cf. Chapter 2.) the bond lengths at the methine atoms of high electron density, i.e. with hybridised orbitals of low *s*-character, at 1.01 Å are longer than those at the methine atoms of low electron density (i.e. orbitals of higher *s*-character) which average at 0.93 Å.

11. References

[1] J. F. LABARRE, F. CRASNIER: Fortschr. chem. Forsch. 24 (1971), 33.
[2] D. LLOYD, D. R. MARSHALL: Angew. Chem. 84 (1972), 447.
[3] Aromaticity Special Publication No. 22, The Chemical Society, Burlington House, London 1967.
[4] E. DALTROZZO, K. FELDMANN: Ber. Bunsenges. physik. Chem. 72 (1968), 1140.
[5] B. BOCK, K. FLATAU, H. JUNGE, M. KUHR, H. MUSSO: Angew. Chem. 83 (1971), 239.
[6] S. DÄHNE, D. LEUPOLD: Ber. Bunsenges. physik. Chem. 70 (1966), 618.
[7] S. DÄHNE, D. LEUPOLD, K.-D. NOLTE: Dye Sensitization: Symposium Bressanone 1967, W. F. BERG, U. MAZZUCATO, M. MEIER and G. SEMERANO: Focal Press, London 1970, 136.
[8] S. DÄHNE: Z. Chem. 10 (1970), 133, 168.
[9] S. DÄHNE: Wiss. Z. Techn. Univ. Dresden 20 (1971), 671.
[10] E. G. COX, D. W. J. CRUICKSHANK, J. A. S. SMITH: Proc. Roy. Soc. (London), Ser. A 247 (1958), 1.
[11] H. EINSPAHR, J. DONOHUE: Acta Crystallogr. B 29 (1973), 1875.
[12] R. D. GILARDI, I. L. KARLE, J. KARLE: Acta Crystallogr. B 28 (1972) 2506; Nature 232 (1971), 187.
[13] J. HJORTAS: Acta Crystallogr. B 28 (1972), 2252.
[14] B. KOCH: Acta Crystallogr. B 28 (1972), 1151.
[15] T. HAMANAKA, T. MITSUI, T. ASHIDA, M. KAKUDO: Acta Crystallogr. B 28 (1972), 214.
[16] H. SCHENK: Acta Crystallogr. B 27 (1971), 667.
[17] P. B. BRAUN, J. HORNSTRA, J. I. LEENHOUTS: Acta Crystallogr. B 27 (1971), 90.
[18] J. F. CHIANG, S. H. BAUER: J. Amer. Chem. Soc. 92 (1970), 261.
[19] C. H. STAM, L. R. DI SANSEVERINO: Acta Crystallogr. 21 (1966), 132.
[20] I. L. KARLE, K. S. DRAGONETTE: Acta Crystallogr. 19 (1965), 500.
[21] W. G. SLY: Acta Crystallogr. 17 (1964), 511.
[22] C. STERLING: Acta Crystallogr. 17 (1964), 500.
[23] R. BIANCHI, A. MUGNOLI, M. SIMONETTA: Acta Crystallogr. B 31 (1975), 1283.
[24] R. BIANCHI, G. CASALONE, M. SIMONETTA: Acta Crystallogr. B 31 (1975), 1207.
[25] A. MUGNOLI, M. SIMONETTA: Acta Crystallogr. B 30 (1974), 2896.
[26] C. M. GRAMACCIOLI, M. SIMONETTA: Acta Crystallogr. B 27 (1971), 2231.
[27] C. M. GRAMACCIOLI, A. MUGNOLI, T. PILATI, M. RAIMONDI, M. SIMONETTA: J. chem. Soc., Chem. Commun. 1971, 973.
[28] C. M. GRAMACCIOLI, A. MIMUN, A. MUGNOLI, M. SIMONETTA: J. chem. Soc., Chem. Commun. 1971, 796.
[29] G. CASALONE, A. GAVEZZOTTI, A. MUGNOLI, M. SIMONETTA: Angew. Chem. 82 (1970), 516.
[30] N. A. BAILEY, R. MASON: Proc. Roy. Soc. (London) Ser. A 290 (1966), 94.
[31] M. DOBLER, J. D. DUNITZ: Helv. chim. Acta 48 (1965), 1429.

[32] J. BREGMAN, F. L. HIRSHFELD, D. RABINOVICH, G. M. J. SCHMIDT: Acta Crystallogr. **19** (1965), 227, 235.
[33] S. M. JOHNSON, I. C. PAUL, G. S. D. KING: J. chem. Soc. (London) **B 1970**, 643.
[34] G. AVITABILE, P. GANIS, V. PETRACCONE: J. physic. Chem. **73** (1969), 2378.
[35] R. BIANCHI, G. MOROSI, A. MUGNOLI, M. SIMONETTA: Acta Crystallogr. **B 29** (1973), 1196.
[36] O. KENNARD, D. L. WAMPLER, J. C. COPPOLA, W. D. S. MOTHERWELL, D. G. WATSON, A. C. LARSON: Acta Crystallogr. **B 27** (1971), 1116.
[37] D. W. HUDSON, O. S. MILLS: J. chem. Soc., Chem. Commun. **1971**, 153.
[38] O. KENNARD, D. G. WATSON, J. K. FAWCETT, K. A. KERR: Tetrahedron (London) **26** (1970), 607.
[39] E. CLAR: Chimia [Aaxau] **18** (1964), 375.
[40] E. CLAR: Polycyclic Hydrocarbons, 2 Vols., Academic Press London and Springer Verlag, Berlin 1964.
[41] J. TROTTER: Lecture Series of the Royal Institute of Chemistry, London 1964, No. 2, p. 1.
[42] H. SOFER, O. E. POLANSKY, G. DERFLINGER: Mh. Chem. **101** (1970), 1318.
[43] H. SOFER, O. E. POLANSKY: Mh. Chem. **102** (1971), 256.
[44] H. SOFER, G. DERFLINGER, O. E. POLANSKY: Mh. Chem. **99** (1968), 1895.
[45] H. SOFER, O. E. POLANSKY, G. DERFLINGER: Mh. Chem. **99** (1968), 1879.
[46] O. E. POLANSKY, G. DERFLINGER: Int. J. Quantum. Chem. **1** (1967), 379.
[47] F. R. AHMED, J. TROTTER: Acta Crystallogr. **16** (1963), 503.
[48] M. J. KAY, Y. OKAYA, D. E. COX: Acta Crystallogr. **B 27** (1971), 26.
[49] A. C. HAZELL, F. K. LARSEN, M. S. LEHMANN: Acta Crystallogr. **B 28** (1972), 2977.
[50] A. CAMERAMAN, J. TROTTER: Proc. Roy. Soc. (London) Ser. **A 279** (1964), 129.
[51] R. THOMAS, P. COPPENS: Acta Crystallogr. **B 28** (1972) 1800; cf. P. T. Cheng, S. C. Nyburg: Acta Crystallogr. **B 29** (1973), 1358.
[52] K. S. DICHMANN, S. C. NYBURG, F. H. PICKARD, J. A. POTWOROWSKI: Acta Crystallogr. **B 30** (1974), 27.
[53] J. H. NOORDIK, T. E. M. VAN DEN HARK, J. J. MOOIJ, A. A. K. KLAASSEN: Acta Crystallogr. **B 30** (1974), 833.
[54] F. A. COTTON, R. EISS: J. Amer. chem. Soc. **91** (1969), 6593.
[55] M. J. BENNETT, F. A. COTTON, P. LEGZDINS: J. Amer. chem. Soc. **90** (1968), 6335.
[56] G. ALLEGRA, A. COLOMBO, A. IMMIRZI, I. W. BASSI: J. Amer. chem. Soc. **90** (1968), 4455.
[57] F. A. COTTON, J. TAKATS: J. Amer. chem. Soc. **90** (1968), 2031.
[58] F. A. COTTON, M. D. LAPRADE: J. Amer. chem. Soc. **90** (1968), 2026.
[59] W. A. C. BROWN, A. T. MCPHAIL, G. A. SIM: J. chem. Soc. (London) **B 1966**, 504.
[60] P. F. LINDLEY, O. S. MILLS: J. chem. Soc. (London) **A 1969**, 1286.
[61] A. DECIAN, R. WEISS: Acta Crystallogr. **B 28** (1972), 3273.
[62] R. EISS: Inorg. Chem. **9** (1970), 1650.
[63] R. MASON, G. B. ROBERTSON: J. chem. Soc. (London) **A 1970**, 1229.
[64] S. KULPE, A. ZEDLER, S. DÄHNE, K.-D. NOLTE: J. prakt. Chem. **315** (1973), 865.
[65] LANDOLT-BÖRNSTEIN. New Series, Group III, Vol. 5. Part a, b. Springer Verlag Berlin, Heidelberg, New York 1971.
[66] O. KENNARD, D. G. WATSON: Crystal Structure Library, Vol I Bibliography of Organic Compounds 1960–1966, University Chemical Laboratory, Cambridge, England 1968.
[67] International Tables for *X*-ray Crystallography, Vols I–III, Kynoch Press, Birmingham, England 1962.
[68] PAULING L.: The Nature of the Chemical Bond, 3rd Edition, Cornell University Press Ithaca, New York, 1960.
[69] A. I. KITAIGORODSKII: Organic Chemical Crystallography, Consultants Bureaus, New York 1961.

[70] A. D. MITCHELL, L. C. CROSS: "Tables of Interatomic Distances and Configuration in Molecules and Ions", Special Publ. No. 11, The Chemical Society, London 1958.
[71] W. A. ISMAILSKY: Dissertation, Techn. Hochschule, Dresden 1913.
[72] S. DÄHNE: Wiss. Z. Techn. Univ. Dresden **22** (1973), 765.
[73] W. KÖNIG: J. prakt. Chem. (2) **112** (1926), 1.
[74] S. DÄHNE: Z. Chem. **5** (1965), 441.
[75] J. FABIAN, H. HARTMANN: J. Signal-A. M. **2** (1974), 457.
[76] J. FABIAN, H. HARTMANN: Theoret. chim. Acta (Berl.) **36** (1975), 351.
[77] J. FABIAN, H. HARTMANN: J. molecular Structure **27** (1975), 67.
[78] E. HEILBRONNER, H. BOCK: Das HMO-Modell und seine Anwendung, Teil I—III, Verlag Chemie, Weinheim/Bergstr. 1968/1970.
[79] A. STREITWIESER: Molecular Orbital Theory for Organic Chemists, J. Wiley and Sons, Inc., Publ. New York, London 1962.
[80] H.-J. RAUH, R. SCHNABEL, G. GEISELER, S. DÄHNE: Z. physik. Chem. (Leipzig) **255** (1974), 651.
[81] S. DÄHNE, H.-J. RAUH, R. SCHNABEL, G. GEISELER: Z. Chem. **13** (1973), 70.
[82] D. LLOYD, D. R. MARSHALL: Chem. Ind. **1972**, 335.
[83] H.-E. NIKOLAJEWSKI, S. DÄHNE, D. LEUPOLD, B. HIRSCH: Tetrahedron (London) **24** (1968), 6685.
[84] S. S. MALHOTRA, M. C. WHITING: J. chem. Soc. (London) **1960**, 3812.
[85] S. DÄHNE, R. RADEGLIA: Tetrahedron (London) **27** (1971), 3673.
[86] S. DÄHNE, K.-D. NOLTE: J. chem. Soc., Chem. Commun. **1972**, 1056.
[87] K.-D. NOLTE, S. DÄHNE: J. prakt. Chem. **318** (1976), 643.
[88] J. KUČERA, Z. ARNOLD: Collect. czechoslov. chem. Commun. **32** (1967), 1704.
[89] H.-E. NIKOLAJEWSKI, S. DÄHNE, B. HIRSCH, E.-A. JAUER: Angew. Chem. **78** (1966), 1063.
[90] S. DÄHNE, S. KULPE, K.-D. NOLTE: Z. Chem. **13** (1973), 422.
[91] S. DÄHNE, S. KULPE, K.-D. NOLTE, R. RADEGLIA: Photogr. Sci. Engng. 18 (1974), 410.
[92] S. NAGAKURA: J. chem. Physics **23** (1955), 1441.
[93] S. NAGAKURA, J. TANAKA: J. chem. Physics **22** (1954), 236.
[94] G. SCHEIBE, W. SEIFFERT, H. WENGENMAYR, C. JUTZ: Ber. Bunsenges. physik. Chem. **67** (1963), 560.
[95] S. DÄHNE, J. RANFT: Z. physik. Chem. (Leipzig) **224** (1963), 65.
[96] R. RADEGLIA, G. ENGELHARDT, E. LIPPMA, T. PEHK, K.-D. NOLTE, S. DÄHNE: Organic Magnetic Resonance **4** (1972), 571.
[97] R. RADEGLIA: J. prakt. Chem. **315** (1973), 1121.
[98] L. MITZINGER: Dissertation, Akademie der Wiss. der DDR, Berlin 1973.
[99] H. H. SEYDEWITZ: Dissertation, Universität Freiburg/Br. 1971.
[100] R. RADEGLIA: J. molecular Structure **17** (1973), 47.
[101] S. DÄHNE, B. BORNOWSKI, B. GRIMM, S. KULPE, D. LEUPOLD, M. NÄTHER: J. Signal-A.M., in press.
[102] S. KULPE, S. DÄHNE, B. ZIEMER, B. SCHULZ: Photogr. Sci. Engng. **20** (1976), 205.
[103] B. GRIMM, S. DÄHNE: J. Signal-A.M. 1 (1973), 339.
[104] H. KOBISCHKE, S. DÄHNE: Photogr. Sci. Engng. **16** (1972), 173.
[105] S. KULPE: Z. Chem. **16** (1976), 56.
[106] S. KULPE: J. prakt. Chem. **316** (1974), 199.
[107] S. KULPE: Tetrahedron (London) **26** (1970), 2899.
[108] S. KULPE: J. prakt. Chem. **312** (1970), 909.
[109] L. PAULING: J. Amer. chem. Soc. **69** (1947), 542.
[110] H.-B. BÜRGI: Angew. Chem. **87** (1975), 461.
[111] M. G. BROWN: Trans. Faraday Soc. **55** (1959), 694.

[112] B. P. Stoicheff: Tetrahedron (London) **17** (1962), 135.
[113] H. J. Bernstein: Trans. Faraday Soc. **57** (1961), 1649.
[114] B. M. Mikhailov: Isvest. Akad. Nauk USSR Otdelenie Khimitcheskij Nauk 8 (1960), 1379.
[115] R. B. Helmholdt, A. F. J. Ruysink, H. Reynaers, G. Kemper: Acta Crystallogr. **B 28** (1972), 318.
[116] T. L. Cottrell: The Strength of the Chemical Bonds, Butterworths, London 1958.
[117] K. Sieber, L. Kutschabsky, S. Kulpe: Kristall u. Technik **9** (1974), 1111.
[118] K. Sieber, L. Kutschabsky, S. Kulpe: Kristall u. Technik **9** (1974), 1101.
[119] K. Sieber: Dissertation, Humboldt-Universität Berlin 1971.
[120] A. Zedler, S. Kulpe: J. prakt. Chem. **317** (1975), 38.
[121] A. Zedler, S. Kulpe: Z. Chem. **10** (1970), 267.
[122] A. Zedler: Dissertation, Akademie der Wiss. der DDR, Berlin 1973.
[123] B. Ziemer, S. Kulpe: J. prakt. Chem. **317** (1975), 185.
[124] B. Ziemer: Dissertation, Akademie der Wiss. der DDR, Berlin 1973.
[125] A. C. Villa, A. G. Manfredotti, M. Nardelli, M. E. V. Tani: Acta Crystallogr. **B 28** (1972), 356.
[126] H. Henke, H. Bärnighausen: Acta Crystallogr. **B 28** (1972), 1100.
[127] J. N. Brown, E. A. Meyers: Acta Crystallogr. **B 26** (1970), 1178.
[128] B. W. Matthews, R. E. Stenkamp, P. M. Colman: Acta Crystallogr. **B 29** (1973), 449.
[129] F. Brandl, H.-J. Springer, P. Narayanan, J. Preuss, W. Hoppe, G. Scheibe: J. Cryst. Mol. Struct. **4** (1974), 391.
[130] J. Kroon, H. Krabbendam: Acta Crystallogr. **B 30** (1974), 1463.
[131] M. Sachs, J. Pletcher, C. S. Yoo, J. M. Stewart: Acta Crystallogr. **B 27** (1971), 1635.
[132] J. Galloy, J.-P. Putzeys, G. Germain, J.-P. Declercq, M. van Meerssche: Acta Crystallogr. **B 30** (1974), 2460.
[133] S. Voliotis, A. Rimsky: Acta Crystallogr. **B 31** (1975), 2612.
[134] S. Voliotis, A. Rimsky, J. Faucherre: Acta Crystallogr. **B 31** (1975), 2607.
[135] I. Kjøller Larsen: Acta Crystallogr. **B 31** (1975), 1626.
[136] M. P. Gupta, B. P. Dutta: Acta Crystallogr. **B 31** (1975), 1272.
[137] L. Coghi, G. Pelizzi: Acta Crystallogr. **B 31** (1975), 131.
[138] J. M. Adams, R. W. H. Small: Acta Crystallogr. **B 30** (1974), 2191.
[139] E. G. Steward, D. Warner, G. R. Clarke: Acta Crystallogr. **B 30** (1974), 813.
[140] C. N. Morimoto, E. C. Lingafelter: Acta Crystallogr. **B 26** (1970), 335.
[141] D. J. Haas, D. R. Harris, H. H. Mills: Acta Crystallogr. **19** (1965), 676.
[142] A. Gieren, F. Pertlik: Acta Crystallogr. **B 31** (1975), 2431.
[143] G. Scheibe, J. Heiss, K. Feldmann: Ber. Bunsenges. physik. Chem. **70** (1966), 52.
[144] H. J. Springer: Dissertation, Technische Universität, München 1965.
[145] P. Goldstein, K. N. Trueblood: Acta Crystallogr. **23** (1967), 148.
[146] H. A. Staab: Einführung in die theoretische organische Chemie, 3. Aufl. Verlag Chemie, Weinheim/Bergstr. 1962.
[147] G. Häfelinger: Chem. Ber. **103** (1970), 2902, 2922, 2941, 3289, 3370.
[148] E. Oeser: Chem. Ber. **107** (1974), 627.
[149] S. Kulpe, B. Schulz: Kristall u. Technik, in preparation.
[150] S. Kulpe, B. Schulz: Z. Chem., in press.
[151] R. Radeglia: J. molecular Structure **17** (1973), 47.
[152] S. Kulpe: Acta Crystallogr. **B 25** (1969), 1411.
[153] G. Scheibe, F. Haimerl, W. Hoppe: Tetrahedron Letters **1970**, 3067.
[154] B. Grimm: Dissertation, Humboldt-Universität, Berlin 1974.
[155] B. Grimm: Z. Chem. **14** (1974), 277.

[156] D. L. Smith: Photogr. Sci. Engng. **18** (**1974**), 309.
[157] P. Gilman: Photochem., Photobiol. **16** (**1972**), 211.
[158] W. West: Photogr. Sci. Engng. **18** (**1974**), 35.
[159] F. Dietz: J. Signal-A.M. **1** (**1973**), 157, 237, 381.
[160] E. Daltrozzo, G. Scheibe, K. Gschwind, F. Haimerl: Photogr. Sci. Engng. 18 (1974), 441.
[161] H. Bücher, H. Kuhn: Chem. Physics Letters **6** (1970), 183.
[162] V. Czikelly, H. D. Försterling, H. Kuhn: Chem. Physics Letters **6** (1970), 11, 207.
[163] G. E. Ficken: Photogr. Sci. Engng. **21** (**1973**), 11.
[164] G. R. Bird: Photogr. Sci. Engng. **15** (**1971**), 134.
[165] K. Norland, A. Ames, T. Taylor: Photogr. Sci. Engng. **14** (1970), 295.
[166] G. R. Bird, B. Zuckerman, A. E. Ames: Photochem. Photobiol. 8 (1968), 393.
[167] G. R. Bird, K. S. Norland, A. E. Rosenoff, H. B. Michaud: Photogr. Sci. Engng. **12** (1968), 196.
[168] F. Dietz, C. Glier: J. Signal-A.M. **2** (**1974**), 123.
[169] F. Dietz: Tetrahedron **28** (**1972**), 1403.
[170] D. L. Smith: Photogr. Sci. Engng. **16** (1972), 329.
[171] S. E. Sheppard: Rev. mod. Physics **14** (**1942**), 303.
[172] H. E. Marr, J. M. Stewart, M. F. Chiu: Acta Crystallogr. **B 29** (1973), 847.
[173] B. Dammeier, W. Hoppe: Acta Crystallogr. **B 27** (1971), 2364.
[174] K. Nakatsu, H. Yoshioka, T. Aoki: Chem. Letters **1972**, 339.
[175] D. L. Smith, E. K. Barrett: Acta Crystallogr. **B 27** (1971), 969.
[176] H. Yoshioka, K. Nakatsu: Chem. Physics Letters **11** (1971), 255.
[177] R. P. Shibaeva, L. O. Atovmyan, V. I. Ponomarjev, O. S. Philipenko, L. P. Rozenberg: Tetrahedron Letters **1973**, 185.
[178] J. Effinger, G. Germain, J. Maunier: Bull. Soc. chim. Belgique **69** (1960), 387.
[179] J. van Thuijl, C. Romers: Acta Crystallogr. **20** (1966), 899.
[180] J. van Thuijl, C. Romers, E. Havinga: Recueil Trav. chim. Pays-Bas **82** (1963), 1026.
[181] H. J. Friedrich, G. Scheibe: Z. Elektrochem., Ber. Bunsenges. physik. Chem. **65** (1961), 851.
[182] G. Scheibe, W. Riess: Chem. Ber. **92** (**1959**), 2189.
[183] J. Potenza, D. Mastropaolo: Acta Crystallogr. **B 30** (1974), 2353.
[184] T. E. Borovjak, N. G. Bokij, Ju. T. Stručkov: Žurnal structurnoj chimii **13** (1972), 480.
[185] D. L. Smith, H. R. Luss: Acta Crystallogr. **B 28** (1972), 2793.
[186] D. L. Smith, H. R. Luss: Acta Crystallogr. **B 31** (1975), 402.
[187] P. J. Wheatley: J. chem. Soc. (London) **1959**, 3245.
[188] P. J. Wheatley: J. chem. Soc. (London) **1959**, 4096.
[189] L. G. Voroncova, Z. V. Zvonkova, G. S. Ždanov: Kristallografija **8** (1963), 374.
[190] W. Falkenberg, E. Höhne, J. Sieler, R. Wagener: Kristall u. Technik 8 (1973), 227.
[191] B. Ziemer, S. Kulpe: J. prakt. Chem. **317** (1975), 199.
[192] L. L. Koh, K. Eriks: Acta Crystallogr. **B 27** (1971), 1405.
[193] G. J. Palenik: Acta Crystallogr. **16** (1963), 596.
[194] K. Britts, I. L. Karle: Acta Crystallogr. **22** (1967), 308.
[195] J. H. Bryden: Acta Crystallogr. **11** (**1958**), 31.
[196] Y. Kushi, Q. Fernando: J. Amer. chem. Soc. **92** (**1970**), **1965**.
[197] Ju. A. Omelčenko, Ju. D. Kondrašev: Kristallografija **16** (1971), 115.
[198] F. P. van Remoortere, F. P. Boer: J. chem. Soc. (London) **B 1971**, **976**.
[199] S. Martinez-Carrera: Acta Crystallogr. **20** (**1966**), **783**.
[200] L. L. Merritt, R. T. Cady, B. W. Mundy: Acta Crystallogr. **7** (1954), 473.
[201] L. Power, J. Pletcher, M. Sax: Acta Crystallogr. **B 26** (**1970**), **143**.

[202] E. Miler-Srenger: Acta Crystallogr. **B 30** (1974), 1911.
[203] E. Miler-Srenger: Acta Crystallogr. **B 29** (1973), 1119.
[204] J. P. Chesick, J. Donohue: Acta Crystallogr. **B 27** (1971), 1441.
[205] P. Stenson: Acta chem. scand. **24** (1970), 3729.
[206] E. Miler-Srenger: Bull. Soc. chim. France **1969**, 3970.
[207] M. Fehlmann: Acta Crystallogr. **B 26** (1970), 1736.
[208] W. Saenger, H. Hettler: Chem. Ber. **102** (1969), 1468.
[209] R. Allmann: Acta Crystallogr. **22** (1967), 246.
[210] J. L. Flippen: Acta Crystallogr. **B 28** (1972), 2749.
[211] T. C. Downie, W. Harrison, E. S. Raper, M. A. Hepworth: Acta Crystallogr. **B 28** (1972), 1584.
[212] G. R. Form, E. S. Raper, T. C. Downie: Acta Crystallogr. **B 31** (1975), 2181.
[213] E. Adman, L. H. Jensen, R. N. Warrener: Acta Crystallogr. **B 31** (1975), 1915.
[214] V. Amirthalingam, K. V. Muralidharan: Acta Crystallogr. **B 28** (1972), 2417, 2421.
[215] J.-P. Mornon, R. Bally: Acta Crystallogr. **B 28** (1972), 2074.
[216] A. F. Cameron, N. J. Hair: J. chem Soc. (London) **B 1971**, 1733.
[217] L. A. Plastas, J. M. Stewart: J. chem. Soc., Chem. Commun. **1969**, 811.
[218] B. W. Matthews: Acta Crystallogr. **17** (1964), 1413.
[219] G. R. Form, E. S. Raper, T. C. Downie: Acta Crystallogr. **B 30** (1974), 342.
[220a] L. K. Hansen, A. Hordvik: Acta chem. scand. **27** (1973), 411.
[220b] F. Leung, S. C. Nyburg: Canad. J. Chem. **49** (1971), 167.
[221] R. Allmann: Chem. Ber. **99** (1966), 1332.
[222] J. J. Daly: J. chem. Soc. (London) **1964**, 3799.
[223] I. Kawada, R. Allmann: Angew. Chem. **80** (1968), 40.
[224] H. Pritzkow, H. Hartl: Acta Crystallogr. **B 29** (1973), 1777.
[225] R. Handa, N. N. Saha: Acta Crystallogr. **B 29** (1973), 554.
[226] P. F. Rodesiler, E. L. Amma: Acta Crystallogr. **B 27** (1971), 1687.
[227] A. Hordvik, S. Joys: Acta chem. scand. **19** (1965), 1539.
[228] A. Hordvik, J. Sletten: Acta chem. scand. **20** (1966), 1907.
[229] U. Müller, H. Bärnighausen: Acta Crystallogr. **B 26** (1970), 1671.
[230] A. Jones, S. Neidle: Acta Crystallogr. **B 31** (1975), 1324.
[231] J. Preuss, A. Gieren, K. Zechmeister, E. Daltrozzo, W. Hoppe, V. Zanker: Chem. Ber. **105** (1972), 203.
[232] H. E. Marr, J. M. Stewart: J. chem. Soc., Chem. Commun. **1971**, 131.
[233] A. Kahn-Harari, R. E. Ballard, E. K. Norris: Acta Crystallogr. **B 29** (1973), 1124.
[234] G. S. Ždanov, Z. V. Zvonkova, L. G. Voroncova: Kristallografija **1** (1956), 61.
[235] H. Taylor: Z. Kristallogr. **A 91** (1935), 450.
[236] S. K. Obendorf, H. L. Carrell, J. Pickworth Glusker: Acta Crystallogr. **B 30** (1974), 1408.
[237] R. Talacki, H. L. Carrell, J. Pickworth Glusker: Acta Crystallogr. **B 30** (1974), 1044.
[238] D. W. Phelps, A. W. Cordes: Acta Crystallogr. **B 31** (1975), 2542.
[239] P. Marsau, J. Gauthier: Acta Crystallogr. **B 29** (1973), 992.
[240] P. Marsau, B. Busetta: Acta Crystallogr. **B 29** (1973), 986.
[241] P. Marsau, Y. Cam: Acta Crystallogr. **B 29** (1973), 980.
[242] S. Dähne, H. Stanko: Naturwissenschaften **50** (1963), 328.
[243] S. Dähne: Z. physik. Chem. (Leipzig) **220** (1962), 187.
[244] F. A. Cotton, B. G. De Boer, J. R. Pipal: Inorg. Chem. **9** (1970), 783.
[245] H. C. Freeman, M. R. Snow: Acta Crystallogr. **18** (1965), 843.
[246] J. E. Johnson, R. A. Jacobson: Acta Crystallogr. **B 29** (1973), 1669.

[247] J. E. JOHNSON, T. A. BEINEKE, R. A. JACOBSON: J. chem. Soc. (London) **A 1971**, 1371.
[248] B. KLEWE: Acta chem. scand. **25** (1971), 1975.
[249] B. KLEWE: Acta chem. scand. **25** (1971), 1988.
[250] J. R. WITT, D. BRITTON: Acta Crystallogr. **B 27** (1971), 1835.
[251] R. DESIDERATO, R. L. SASS: Acta Crystallogr. 18 (1965), 1.
[252] J. EDMONDS, J. K. HERDKLOTZ, R. L. SASS: Acta Crystallogr. **B 26** (1970), 1355.
[253] D. BRITTON: Acta Crystallogr. **B 30** (1974), 1818.
[254] A. E. DOUGLAS, D. SHARMA: J. chem. Physics 21 (1953), 448.
[255a] F. E. MORRIS, W. J. ORVILLE-THOMAS: J. molecular Structure **6** (1961), 572.
[255b] T. HAHN: Z. Kristallogr. **109** (1957), 438.
[256] P. L. MARKILA, S. J. RETTIG, J. TROTTER: Acta Crystallogr. **B 31** (1975), 2927.
[257] J. C. SPEAKMAN, H. H. MILLS: J. chem. Soc. (London) **1961**, 1164.
[258] E. J. GABE, M. R. TAYLOR: Acta Crystallogr. **21** (1966), 418.
[259a] J. H. ROBERTSON: Acta Crystallogr. **18** (1965), 410.
[259b] B. BEAGLEY, R. W. H. SMALL: Acta Crystallogr. **17** (1964), 783.
[260] M. N. G. JAMES, G. J. B. WILLIAMS: Acta Crystallogr. **B 30** (1974), 1257.
[261] S. S. TAVALE, L. M. PANT, A. B. BISWAS: Acta Crystallogr. **14** (1961), 1281.
[262] B. BEAGLEY, W. H. SMALL: Proc. Roy. Soc. (London) Ser. **A 275** (1963), 469.
[263] D. W. J. CRUICKSHANK, D. W. JONES, G. WALKER: J. chem. Soc. (London) **1964**, 1303.
[264] S. S. TAVALE, L. M. PANT, A. B. BISWAS: Acta Crystallogr. **16** (1963), 566.
[265] A. GIEREN, W. HOPPE, K. FLEISCHMANN: Angew. Chem. **85** (1973), 308.
[266] J. M. ADAMS, R. W. SMALL: Acta Crystallogr. **B 29** (1973), 2317.
[267] D. W. THOMPSON: Structure and Bonding **9** (1971), 27.
[268] F. BONATI: Organometallic Chem. Rev. **1** (1966), 379.
[269] E. C. LINGAFELTER: Coordinat. Chem. Rev. **1** (1966), 151.
[270] J. SELBIN: Chem. Reviews **65** (1965), 153.
[271] F. A. SCHRÖDER, H. P. WEBER: Acta Crystallogr. **B 31** (1975), 1745.
[272] J. M. STEWART, B. MOROSIN: Acta Crystallogr. **B 31** (1975), 1164.
[273] B. W. MCCLELLAND: Acta Crystallogr. **B 31** (1975), 2496.
[274] K. DYMOCK, G. J. PALENIK: Acta Crystallogr. **B 30** (1974), 1364.
[275] G. W. BUSHNELL: Canad. J. Chem. **49** (1971), 555.
[276] T. PHILLIPS, D. E. SANDS, W. F. WAGNER: Inorg. Chem. **7** (1968), 2295.
[277] B. MOROSIN: Acta Crystallogr. **19** (1965), 131.
[278] C. E. WILKES, R. A. JACOBSON: Inorg. Chem. **4** (1965), 99.
[279] M. D. GLICK, S. R. B. ANDRELCZYK, R. L. LINTVEDT: Acta Crystallogr. **B 31** (1975), 916.
[280] J. IBALL, C. H. MORGAN: Acta Crystallogr. **23** (1967), 239.
[281] F. A. COTTON, R. EISS: J. Amer. chem. Soc. **90** (1968), 38.
[282] G. J. KRUGER, E. C. REYNHARDT: Acta Crystallogr. **B 30** (1974), 822.
[283] W. D. HORROCKS, D. H. TEMPLETON, A. ZALKIN: Inorg. Chem. **7** (1968), 1552.
[284] G. KUNG-JOU CHAO, R. L. SIME, R. J. SIME: Acta Crystallogr. **B 29** (1973), 2845.
[285] R. E. CRAMER, K. SEFF: Acta Crystallogr. **B 28** (1972), 3281.
[286] J. A. CUNNINGHAM, D. E. SANDS, W. F. WAGNER, M. F. RICHARDSON: Inorg. Chem. **8** (1969), 22.
[287] E. D. WATKINS, J. A. CUNNINGHAM, T. PHILLIPS, D. E. SANDS, W. F. WAGNER: Inorg. Chem. **8** (1969), 29.
[288] M. F. RICHARDSON, P. W. R. CORFIELD, D. E. SANDS, R. E. SIEVERS: Inorg. Chem. **9** (1970), 1632.
[289] E. C. LINGAFELTER, R. L. BRAUN: J. Amer. chem. Soc. **88** (1966), 2951.
[290] J. P. COLLMAN: Angew. Chem. **77** (1965), 154.
[291] M. CALVIN, K. W. WILSON: J. Amer. chem. Soc. **67** (1945), 2003.

[292] R. C. FAY, N. SEPONE: J. Amer. chem. Soc. **90** (1968), 5701.
[293] N. SEPONE, R. C. FAY: Inorg. Chem. 8 (1969), 2379.
[294] S. TACHIYASHIKI, H. NAKAYAMA, R. KURODA, S. SATO, Y. SAITO: Acta Crystallogr. **B 31** (1975), 1483.
[295] M. ELDER: Inorg. Chem. 8 (1969), 2103.
[296] F. A. COTTON, B. H. C. WINQUIST: Inorg. Chem. 8 (1969), 1304.
[297] M. BLACKSTONE, J. v. THUIJL, C. ROMERS: Recueil Trav. chim. Pays-Bas **85** (1966), 557.
[298] F. J. HOLLANDER, D. H. TEMPLETON, A. ZALKIN: Acta Crystallogr. **B 29** (1973), 1289, 1295, 1303.
[299] N. V. GRIGOREVA, N. V. MARGOLIS, I. N. ŽOCHOR, V. V. MELNIKOV, I. V. CELINSKIJ: Žurnal strukturnoj chimii **7** (1966), 278.
[300] N. V. GRIGOREVA, N. V. MARGOLIS, I. V. CELINSKIJ, V. V. MELNIKOV, G. V. MAKARENKO: Žurnal strukturnoj chimii **12** (1971), 938.
[301] H. UEDA, N. SAKABE, J. TANAKA, A. FURUSAKI: Bull. chem. Soc. Japan **41** (1968), 2866.
[302] R. DESTRO, C. M. GRAMACCIOLI, M. SIMONETTA: Acta Crystallogr. **B 24** (1968), 1369.
[303] G. G. MESSMER, G. J. PALENIK: J. chem. Soc., Chem. Commun. **1969**, 470.
[304] M. MATHEW, G. J. PALENIK: Acta Crystallogr. **B 27** (1971), 1388.
[305] C. K. PROUT, O. J. R. HODDER, D. VITERBO: Acta Crystallogr. **B 28** (1972), 1523.
[306] R. SHIONO: Acta Crystallogr. **14** (1961), 42.
[307] E. L. MUETTERTIES, L. J. GUGGENBERGER: J. Amer. chem. Soc. **94** (1972), 8046.
[308] J. J. PARK, D. M. COLLINS, J. L. HOARD: J. Amer. chem. Soc. **92** (1970), 3636.
[309] T. A. HAMOR, D. J. WATKIN: J. chem. Soc., Chem. Commun. **1969**, 440.
[310] R. J. IRVING, M. L. POST, D. C. POVEY: J. chem. Soc., Dalton Transact. **1973**, 697.
[311] V. W. DAY, J. L. HOARD: J. Amer. chem. Soc. **92** (1970), 3626.
[312] G. P. KHARE, A. J. SCHULTZ, R. EISENBERG: J. Amer. chem. Soc. **93** (1971), 3597.
[313] Y. OTAKA, F. MARUMO, Y. SAITO: Acta Crystallogr. **B 28** (1972), 1590.
[314] A. HORDVIK, H. M. KJÖGE: Acta chem. scand. **20** (1966), 1923.
[315] A. HORDVIK, E. SLETTEN: Acta chem. scand. **20** (1966), 2043.
[316] L. FANFANI, A. NUNZI, P. F. ZANAZZI, A. R. ZANZARI: Acta Crystallogr. **B 28** (1972), 1298.
[317] M. BONAMICO, G. DESSY, V. FARES, L. SCARAMUZZA: J. chem. Soc., Dalton Transact. **1972**, 2515.
[318] G. M. SHELDRICK, W. S. SHELDRICK: J. chem. Soc. (London) **A 1970**, 490, 493.
[319] F. W. B. EINSTEIN, J. S. FIELD: Acta Crystallogr. **B 30** (1974), 2928.
[320] K. A. FRASER, M. M. HARDING: Acta Crystallogr. **22** (1967), 75.
[321] H. P. KLUG: Acta Crystallogr. **21** (1966), 536.
[322] G. R. DAVIES, J. A. J. JARVIS, B. T. KILBOURN, R. H. B. MAIS, P. G. OWSTON: J. chem. Soc. (London) **A 1970**, 1275.
[323] P. W. G. NEWMAN, A. H. WHITE: J. chem. Soc., Dalton Transact. **1972**, 1460.
[324] J. POTENZA, D. MASTROPAOLO: Acta Crystallogr. **B 29** (1973), 1830.
[325] C. S. HARRELD, E. O. SCHLEMPER: Acta Crystallogr. **B 27** (1971), 1964.
[326] H. IWASAKI, H. HAGIHARA: Acta Crystallogr. **B 28** (1972), 507.
[327] C. FABIANI, R. SPAGNA, A. VACIAGO, L. ZAMBONELLI: Acta Crystallogr. **B 27** (1971), 1499.
[328] M. BONAMICO, A. VACIAGO, L. ZAMBONELLI: Acta Crystallogr. **19** (1965), 619, 886, 898.
[329] A. DOMENICANO, L. TORELLI, A. VACIAGO, L. ZAMBONELLI: J. chem. Soc. (London) **A 1968**, 1351.
[330] P. C. HEALY A. H. WHITE: J. chem. Soc., Dalton Transact. **1972**, 1883.
[331] S. MERLINO: Acta Crystallogr. **B 24** (1968), 1441.
[332] V. M. AGRE, E. A. ŠUGAM: Žurnal strukturnoj chimii **13** (1972), 660.

[333] G. Peyronel, A. Pignedoli, L. Antolini: Acta Crystallogr. **B 28** (1970), 3596.

[334] A. Chiesi Villa, A. Gaetani Manfredotti, C. Guastini, M. Nardelli: Acta Crystallogr. **B 28** (1972), 2231.

[335] H. C. Freeman, G. H. W. Milburn, C. E. Nockolds, R. Mason, G. B. Robertson, G. A. Rusholme: Acta Crystallogr. **B 30** (1974), 886.

[336] R. Mason, G. B. Robertson, G. A. Rusholme: Acta Crystallogr. **B 30** (1974), 906.

[337] R. Mason, G. B. Robertson, G. A. Rusholme: Acta Crystallogr. **B 30** (1974), 894.

[338] F. Grundtvig, A. Hordvik: Acta chem scand. **25** (1971), 1567.

[339] A. Hordvik, H. M. Kjöge: Acta chem. scand. **23** (1969), 1367.

[340] A. Hordvik, R. M. Baxter: Acta chem. scand. **23** (1969), 1082.

[341] A. Hordvik, E. Sletten: Acta chem. scand. **20** (1966), 1874.

[342] A. Hordvik, H. M. Kjöge: Acta chem. scand. **19** (1965), 935.

[343] A. Hordvik: Acta chem. scand. **19** (1965), 1039.

[344] B. Duffin: Acta Crystallogr. **B 24** (1968), 1256.

[345] S. Dähne, F. Schob, K.-D. Nolte: Z. Chem. **13** (1973), 471.

[346] R. Radeglia: Z. physik. Chem. (Leipzig) **235** (1967), 335.

[347] R. Radeglia, S. Dähne: Ber. Bunsenges. physik. Chem. **70** (1966), 745.

[348] K.-D. Nolte: Dissertation, Deutsche Akademie der Wiss., Berlin 1972; K.-D. Nolte, S. Dähne: J. prakt. Chem. **318** (1976) 993.

[349] R. Radeglia, S. Dähne: J. molecular Structure **5** (1970), 399.

[350] S. Dähne, D. Leupold, H.-E. Nikolajewski, R. Radeglia: Z. Naturforsch. **20b** (1965), 1006.

[351] S. Kulpe, B. Schulz: Kristall u. Technik, in preparation.

[352] S. Kulpe, B. Schulz: Kristall und Technik, **11** (1976), 707.

[353] S. Kulpe: Z. Chem., **16** (1976), 153.

[354] M. Klessinger: Theoret. chim. Acta (Berl.) **5** (1966), 251.

[355] D. Leupold, S. Dähne: Theoret. chim. Acta (Berl.) **3** (1965), 1.

[356] S. Kulpe, B. Schulz: Z. Chem., in press.

[357] K.-D. Nolte, S. Dähne: J. Mol. Relax. Processes, in press.

[358] W. Hamilton: Acta Crystallogr. **18** (1965), 866.

[359] D. O. Hughes, R. W. H. Small: Acta Crystallogr. **15** (1962), 933.

[360] P. C. Chieh, E. Subramanian, J. Trotter: J. chem. Soc. (London) **A 1970**, 179.

[361] M. Hospital, J. Housty: Acta Crystallogr. **21** (1966), 413.

[362] M. Hospital, J. Housty: Acta Crystallogr. **20** (1966), 626.

[363] M. Hospital, J. Housty: Acta Crystallogr. **20** (1966), 368.

[364] M. Hospital: Acta Crystallogr. **B 27** (1971), 484.

[365] A. Grand, C. Cohen-Addad: Acta Crystallogr. **B 29** (1973), 1149.

[366] R. B. Corey, J. Donohue: J. Amer. chem. Soc. **72** (1950), 2899.

[367] J. L. Katz, B. Post: Acta Crystallogr. **13** (1960), 624.

[368] P. Groth: Acta chem. scand. **23** (1969), 3155.

[369] D. G. R. Yeates, R. A. Palmer: Acta Crystallogr. **B 31** (1975), 1077.

[370] R. Mason: Acta Crystallogr. **14** (1961), 720.

[371] G. Argay, J. Seres: Acta Crystallogr. **B 29** (1973), 1146.

[372] G. Argay, A. Kalman: Acta Crystallogr. **B 29** (1973), 636.

[373] A. Kirfel: Acta Crystallogr. **B 31** (1975), 2494.

[374] A. Kirfel, G. Will, K. Fickentscher: Acta Crystallogr. **B 31** (1975), 1973.

[375] G. S. D. King: J. chem. Soc. (London) **B 1966**, 1224.

[376] E. Höhne, I. Seidel, H. Schmittler-Fichtner, P. Leibnitz: Kristall u. Technik 8 (1973), 207.

[377] W. R. Krigbaum, R.-J. Roe, J. D. Woods: Acta Crystallogr. **B 24** (1968), 1304.

[378] P. VAUGHAN, J. DONOHUE: Acta Crystallogr. **5** (1952), 530.
[379] P. GANIS, G. AVITABILE, E. BENEDETTI, C. PEDONE, M. GOODMAN: Proc. nat. Acad. Sci. USA **67** (1970), 426.
[380] M. NARDELLI, G. F. GASPARRI, P. BOLDRINI, G. G. BATTISTINI: Acta Crystallogr. **19** (1965), 491.
[381] S. C. CHOI, E. PRINCE: Acta Crystallogr. **B 28** (1972), 2857.
[382] A. SANTORO, C. S. CHOI, J. E. ABEL: Acta Crystallogr. **B 31** (1975), 2126.
[383] J. W. TURLEY: Acta Crystallogr. **B 24** (1968), 942.
[384] R. HUBER, R. LANGER, W. HOPPE: Acta Crystallogr. **18** (1965), 467.
[385] M. R. TRUTER: J. chem. Soc. (London) **1960**, 997.
[386] R. L. GIRLING, J. E. O'CONNOR, E. L. AMMA: Acta Crystallogr. **B 28** (1972), 2640.
[387] W. WALTER, J. HOLST: Naturwissenschaften **56** (1969), 327.
[388] A. PIGNEDOLI, G. PEYRONEL, L. ANTOLINI: Acta Crystallogr. **B 31** (1975), 1903.
[389] A. PIGNEDOLI, G. PEYRONEL, L. ANTOLINI: Acta Crystallogr. **B 29** (1973), 1490.
[390] H. LUTH, E. A. HALL, W. A. SPOFFORD, E. L. AMMA: J. chem. Soc., Chem. Commun. **1969**, 520.
[391] R. L. GIRLING, E. L. AMMA: J. chem. Soc., Chem. Commun. **1968**, 1487.
[392] M. R. TRUTER: Acta Crystallogr. **22** (1967), 556.
[393] Z. V. ZVONKOVA, V. J. KRIVNOV, A. N. CHVATKINA: Kristallografija **11** (1966), 385.
[394] W. A. SPOFFORD, E. A. H. GRIFFITH, E. L. AMMA: J. chem. Soc., Chem. Commun. **1970**, 533.
[395] E. A. VIZZINI, J. F. TAYLOR, E. L. AMMA: Inorg. Chem. **7** (1968), 1351.
[396] J. E. O'CONNOR, E. L. AMMA: Inorg. Chem. **8** (1969), 2367.
[397] H. LUTH, M. R. TRUTER: J. Chem. Soc. (London) **A 1968**, 1879.
[398] L. M. AMZEL, S. BAGGIO, L. N. BECKA: J. chem. Soc. (London) **A 1969**, 2066.
[399] D. A. BERTA, W. A. SPOFFORD, P. BOLDRINI, E. L. AMMA: Inorg. Chem. **9** (1970), 136.
[400] G. D. ANDREETTI, P. DOMIANO, G. F. GASPARRI, M. NARDELLI, P. SGARABOTTO: Acta Crystallogr. **B 26** (1970), 1005.
[401] F. HANSEN, R. GROENBAEK-HAZELL: Acta chem. scand. **23** (1969), 1359.
[402] R. E. BALLARD, D. B. POWELL. U. A. JAYASOORIYA: Acta Crystallogr. **B 30** (1974), 1111.
[403] G. HUNG-YIN-LIN, H. HOPE: Acta Crystallogr. **B 28** (1972), 643.
[404] E. LARSEN, P. TRINDERUP: Acta chem. scand. **24** (1970), 261.
[405] M. MATHEW, G. J. PALENIK: J. Amer. chem. Soc. **91** (1969), 6310.
[406] V. AMIRTHALINGAM, V. S. JAKKAL: Acta Crystallogr. **B 28** (1972), 2612.
[407] Y. M. CHOW. Inorg. Chem. **9** (1970), 794.
[408] J. DONOHUE: Acta Crystallogr. **B 24** (1968), 1558.
[409] E. K. ANDERSEN, I. G. K. ANDERSEN: Acta Crystallogr. **B 31** (1975), 394.
[410] J. L. LAWRENCE, S. G. G. MACDONALD: Acta Crystallogr. **B 25** (1969), 1247.
[411] J. HVOSLEF: Acta Crystallogr. **B 24** (1968), 23.
[412] R. ALLMANN, A. KUTOGLU: Acta Crystallogr. **B 31** (1975), 2734.
[413] D. SEMMINGSEN: Acta chem. scand. **B 28** (1974), 141.
[414] F. J. HOLLANDER, D. H. TEMPLETON, A. ZALKIN: Acta Crystallogr. **B 29** (1973), 1552.
[415] D. E. WILLIAMS: Acta Crystallogr. **21** (1966), 340.
[416] R. MASON, G. B. ROBERTSON, P. J. PAULING: J. chem. Soc. (London) **A 1969**, 485.
[417] J. W. CARMICHAEL, L. K. STEINRAUF, R. L. BELFORD: J. chem. Physics **43** (1965), 3959.
[418] P.-K. HON, R. L. BELFORD, C. E. PFLUGER: J. chem. Physics **43** (1965), 1323, 3111.
[419] P.-K. HON, C. E. PFLUGER, R. L. BELFORD: Inorg. Chem. **5** (1966), 516.
[420] R. L. BELFORD, N. D. CHASTEEN, M. A. HITCHMAN, P.-K. HON, C. E. PFLUGER, I. C. PAUL: Inorg. Chem. **8** (1969), 1312.

[421] B. Kojić-Prodić, Ž. Ružic-Toroš, D. Grdenić, L. Golič: Acta Crystallogr. **B 30** (1974), 300.
[422] G. A. Barclay, A. Cooper: J. chem. Soc. (London) **1965**, 3746.
[423] D. Bruins, D. L. Weaver: Inorg. Chem. **9** (1970), 130.
[424] G. R. Clark, D. Hall, T. N. Waters: J. chem. Soc. (London) **A 1968**, 223.
[425] S. R. Holbrook, D. van der Helm: Acta Crystallogr. **B 31** (1975), 1653.
[426] J. Silverman, N. F. Yannon: Acta Crystallogr. **18** (1965), 756.
[427] S. Kulpe, B. Schulz: Kristall u. Technik, in preparation.
[428] L. M. Školnikova, Ju. M. Jutal, E. A. Šugam, A. N. Knjazeva: Žurnal strukturnoj chimii **14** (1973), 92.
[429] J. J. de Boer, D. Bright: Acta Crystallogr. **B 31** (1975), 2342.
[430] A. Hordvik: Acta chem. scand. **20** (1966), 754.
[431] J. F. McConnell, M. J. Lacey, C. G. Macdonald, J. S. Shannon: Acta Crystallogr. **B 29** (1973), 2477.
[432] D. Carlström: Acta chem. scand. **20** (1966), 1240.
[433] Ju. A. Omelčenko, Ju. D. Kondrašev, S. L. Ginzburg, M. G. Neuhaus: Kristallografija **19** (1974), 522.
[434] J. Preuss, A. Gieren: Acta Crystallogr. **B 31** (1975), 1276.
[435] B. Klewe: Acta chem. scand. **25** (1971), 1999.
[436] J. Delettré, R. Bally, J.-P. Mornon: Acta Crystallogr. **B 31** (1975), 2117.
[437] J. L. Flippen, I. L. Karle: J. physic. Chem. **74**, (1970), 769.
[438] R. H. Stanford: Acta Crystallogr. **16** (1963), 1157.
[439] A. Hordvik: Acta chem. scand. **17** (1963), 2575; **15** (1961), 1186.
[440a] G. Germain, P. Piret, M. van Meerssche, J. de Kerf: Bull. Soc. chim. France **1961**, 1407.
[440b] M. Bois D'Enghien-Peteau, J. Meunier-Piret, M. van Meerssche: J. Chim. physique **65** (1968), 1221.
[440c] G. Germain, C. Paternotte, P. Piret, M. van Meerssche: J. Chim. physique **61** (1964), 1059.
[440d] G. Germain, P. Piret, M. van Meerssche: Acta Crystallogr. **15** (1962), 373.
[440e] D. van der Helm, A. E. Lessor, L. L. Merritt: Acta Crystallogr. **15** (1962), 1227.
[440f] H. R. Luss, D. L. Smith: Acta Crystallogr. **B 29** (1973), 998.
[441] J. J. Daly: J. chem. Soc. (London) **1961**, 2801.
[442] E. Schaumann, A. Röhr, S. Sieveking, W. Walter: Angew. Chem. **87** (1975), 486.
[443] R. Radeglia, E. Gey, G. W. Fischer: Z. Chem. **10** (1970), 300.
[444] G. Scheibe, W. Seiffert, G. Hohlneicher, C. Jutz, H.-J. Springer: Tetrahedron Letters **1966**, 5053.
[445] J. R. Holden, C. Dickinson: J. Amer. chem. Soc. **90** (1968), 1975.
[446] T. S. Cameron, C. K. Prout: J. chem. Soc. (London) **C 1969**, 2285.
[447] T. S. Cameron, C. K. Prout: J. chem. Soc. (London) **C 1969**, 2289.
[448] T. S. Cameron, C. K. Prout: J. chem. Soc. (London) **C 1969**, 2292.
[449] A. S. Bailey, J. M. Peach, T. S. Cameron, C. K. Prout: J. chem. Soc. (London) **C 1969**, 2295.
[450] C. Foces-Foces, F. H. Cano, S. Garcia-Blanco, J. G. Rodriguez: Acta Crystallogr. **B 31** (1975), 2310.
[451] A. S. Bailey, W. A. Warr, G. B. Allison, C. K. Prout: J. chem. Soc. (London) **C 1970**, 956.
[452] L. Leiserowitz, D. Rabinovich: J. chem. Soc. (London) **A 1969**, 2367.
[453] T. S. Cameron, D. J. Cowley, J. E. Thompson: J. chem. Soc., Perkin Transact. II **1974**, 774.
[454] A. W. Hanson: Acta Crystallogr. **B 31** (1975), 1963.

[455] S. Shimizu, S. Kashino, M. Haisa: Acta Crystallogr. **B 31** (1975), 1287.
[456] J. C. Messager, Y. Delugeard: Cryst. Struct. Commun. **3** (1974), 391.
[457] M. S. Shen, R. F. Bryan: Acta Crystallogr. **B 31** (1975), 2907.
[458] R. S. Osborn, D. Rogers: Acta Crystallogr. **B 31** (1975), 359.
[459] L. Fallon, H. L. Ammon, R. West, V. N. M. Rao: Acta Crystallogr. **B 30** (1974), 2407.
[460] H. L. Ammon: Acta Crystallogr. **B 30** (1974), 1731.
[461] P. A. Baron, R. D. Brown, R. F. Burden, P. J. Domaille, J. E. Kent: J. molecular Spectroscopy **43** (1972), 401.
[462] H. L. Ammon, G. L. Wheeler: J. chem. Soc., Chem. Commun. **1971**, 1032.
[463] H. L. Ammon, L. A. Plastas: J. chem. Soc., Chem. Commun. **1971**, 356.
[464] H. Burzlaff, K. Hartke, R. Salomon: Chem. Ber. **103** (1970), 156.
[465] H. Burzlaff, I. Kawada: Chem. Ber. **102** (1969), 3632.
[466a] N. Norman, B. Post: Acta Crystallogr. **14** (1961), 503.
[466b] R. Böhme, H. Burzlaff: Chem. Ber. **107** (1974), 832.
[467] M. D. Danford, R. L. Livingston: J. Amer. chem. Soc. **77** (1955), 2944.
[468] L. F. Thomas, E. I. Sherrard, J. Sheridan: Trans. Faraday Soc. **51** (1955), 619.
[469] H. Shimanouchi, Y. Sasada: Acta Crystallogr. **B 29** (1973), 81.
[470] T. A. Hamor, J. E. Derry: Acta Crystallogr. **B 29** (1973), 2649.
[471] D. W. J. Cruickshank, G. Filippini, O. S. Mills: J. chem. Soc., Chem. Commun. **1972**, 101.
[472] D. J. Watkin, T. A. Hamor: J. chem. Soc. (London) **B 1971**, 2167.
[473] K. Ibata, H. Shimanouchi, Y. Sasada, T. Hata: Acta Crystallogr. **B 31** (1975), 2313.
[474] K. Ibata, H. Shimanouchi, Y. Sasada: Acta Crystallogr. **B 31** (1975), 482.
[475] H. Shimanouchi, Y. Sasada, C. Kabuto, Y. Kitahara: Acta Crystallogr. **B 30** (1974), 1273.
[476] H. Shimanouchi, Y. Sasada, C. Kabuto, Y. Kitahara: Acta Crystallogr. **B 30** (1974), 1267.
[477] A. Hoekstra, P. Meertens, A. Vos: Acta Crystallogr. **B 31** (1975), 2813.
[478] K. Plieth, H. Repmann: Z. Kristallogr. **124** (1967), 77.
[479] C. J. Brown: Acta Crystallogr. **21** (1966), 153.
[480] H. Hope, D. Victor: Acta Crystallogr. **B 25** (1969), 1849.
[481] J. Bernstein: Acta Crystallogr. **B 31** (1975), 1268.
[482] C. J. Finder, M. G. Newton, N. L. Allinger: Acta Crystallogr. **B 30** (1974), 411.
[483] L. A. Četkina, G. A. Golder: Žurnal strukturnoj chimii **9** ((1968), 250.
[484] G. Ruban, P. Luger: Acta Crystallogr. **B 31** (1975), 2658.
[485] E. J. Graeber, B. Morosin: Acta Crystallogr. **B 30** (1974), 310.
[486] J. P. Chesick: Acta Crystallogr. **B 29** (1973), 2309.
[487] A. T. Vincent, P. J. Wheatley: J. chem. Soc., Perkin Transact. II **1972**, 1567.
[488] W. R. Krigbaum, P. G. Barber: Acta Crystallogr. **B 27** (1971), 1884.
[489] Y. Sasada, H. Shimanouchi, H. Nakanishi, M. Hasegawa: Bull. chem. Soc. Japan **44** (1971), 1262.
[490] U. C. Sinha: Acta Crystallogr. **B 26** (1970), 889.
[491] J. G. Leipoldt, S. S. Basson, L. D. C. Bok: Acta Crystallogr. **B 26** (1970), 361.
[492] W. R. Krigbaum, Y. Chatani, P. G. Barber: Acta Crystallogr. **B 26** (1970), 97.
[493] C. J. Brown: Acta Crystallogr. **21** (1966), 146.
[494] M. van Meerssche, G. Leroy: Bull. Soc. chim. Belgique **69** (1960), 204.
[495] H. B. Bürgi, J. D. Dunitz: Helv. chim. Acta **53** (1970), 1747.
[496] D. P. Lesser, A. de Vries, J. W. Reed, G. H. Brown: Acta Crystallogr. **B 31** (1975), 653.
[497] D. L. Smith, E. K. Barrett: Acta Crystallogr. **B 27** (1971), 2043.
[498] A. W. Hanson: Acta Crystallogr. **B 29** (1973), 454.

[499] D. Rabinovich: J. chem. Soc. (London) **A 1969, 2361.**
[500] F. H. Allen, J. Trotter, D. Rogers: J. chem. Soc. (London) **B 1971, 166.**
[501] H. A. Fraterman, C. Romers: Recueil Trav. chim. Pays-Bas **90** (1971), 364.
[502] C. Romers, B. Hesper: Acta Crystallogr. **20** (1966), 162.
[503] C. T. Presley, R. L. Sass: Acta Crystallogr. **B 26** (1970), 1195.
[504] C. Römming: Acta chem. scand. **17** (1963), 1444.
[505] Ja. M. Nesterova, M. A. Poraj-Košic: Žurnal strukturnoj chimii **12** (1971), 108.
[506] R. Allmann: Z. Kristallogr. **128** (1969), 115.
[507] C. Reichardt, K. Dimroth: Fortschr. chem. Forsch. **11** (1968), 1.
[508] W. Liptay: Angew. Chem. **81** (1969), 195.
[509] L. V. Vilkov, T. P. Timaševa: Dokl. Akad. Nauk **161** (1965), 351.
[510] J. K. Dattagupta, N. N. Saha: Acta Crystallogr. **B 29** (1973), 1228.
[511] A. K. Pant: Acta Crystallogr. **19** (1965), 440.
[512] K. N. Trueblood, E. Goldish, J. Donohue: Acta Crystallogr. **14** (1961), 1009.
[513] T. C. W. Mak, J. Trotter: Acta Crystallogr. **18** (1965), 68.
[514] J. R. Holden, C. Dickinson, C. M. Bock: J. physic. Chem. **76** (1972), 3597.
[515] J. V. Barve, L. M. Pant: Acta Crystallogr. **B 27** (1971), 1158.
[516] P. Coppens: Acta Crystallogr. **17** (1964), 573.
[517] P. Coppens, G. M. J. Schmidt: Acta Crystallogr. **17** (1964), 222.
[518] J. Trotter: Tetrahedron (London) **8** (1960), 13.
[519] T. D. Sakore, L. M. Pant: Acta Crystallogr. **21** (1966), 715.
[520] Chang Sun Choi, J. E. Abel: Acta Crystallogr. **B 28** (1972), 193.
[521] J. R. Holden: Acta Crystallogr. **22** (1967), 545.
[522] H. H. Cady, A. C. Larson: Acta Crystallogr. **18** (1965), 485.
[523] G. G. Christoph, E. B. Fleischer: Acta Crystallogr. **B 29** (1973), 121.
[524] P. Batail, D. Grandjean, F. Dudragne, C. Michaud: Acta Crystallogr. **B 31** (1975), 1367.
[525] P. Batail, D. Grandjean, F. Dudragne, C. Michaud: Acta Crystallogr. **B 30** (1974), 2653.
[526] J. R. Holden, C. Dickinson: J. physic. Chem. **73** (1969), 1199.
[527] H. H. Cady: Acta Crystallogr. **23** (1967), 601.
[528] T. S. Cameron, C. K. Prout: J. chem. Soc. (London) **C 1969**, 2281.
[529] E. K. Andersen, I. G. K. Andersen: Acta Crystallogr. **B 31** (1975), 387.
[530] E. K. Andersen, I. G. K. Andersen: Acta Crystallogr. **B 31** (1975), 391.
[531] R. J. Neustadt, F. W. Cagle, jr.: Acta Crystallogr. **B 31** (1975), 2727.
[532] E. Carstensen-Oeser, S. Göttlicher, G. Habermehl: Chem. Ber. **101** (1968), 1648.
[533] P. Coppens, G. M. J. Schmidt: Acta Crystallogr. **18** (1965), 62, 654.
[534] J. P. G. Richards: Z. Kristallogr. **116** (1961), 468.
[535] F. Pandarese, L. Ungaretti, A. Coda: Acta Crystallogr. **B 31** (1975), 2671.
[536] A. Chiaroni: Acta Crystallogr. **B 27** (1971), 448.
[537] Y. Delugeard, J. C. Messager: Acta Crystallogr. **B 31** (1975), 2809.
[538] G. Menczel: Acta chim. Acad. Sci. hung. **62** (1969), 41.
[539] C. E. Pfluger, R. L. Harlow: Acta Crystallogr. **B 29** (1973), 2608.
[540] G. Arcovito, M. Bonamico, A. Domenicano, A. Vaciago: J. chem. Soc. (London) **B 1969**, 733.
[541] G. Marongiu, E. C. Lingafelter: Acta Crystallogr. **B 27** (1971), 1195.
[542] P. L. Orioli, M. Di Vaira, L. Sacconi: Inorg. Chem. **5** (1966), 400.
[543] K. Tsukuma, T. Kawaguchi, T. Watanabé: Acta Crystallogr. **B 31** (1975), 2165.
[544] G. Gilli, D. W. J. Cruickshank, R. L. Beddoes, O. S. Mills: Acta Crystallogr. **B 28** (1972), 1889.

[545] M. Calligaris, G. Nardin, L. Randaccio: J. chem. Soc., Dalton Transact. **1972**, 1433.
[546] B. M. Gatehouse, B. E. Reichert, B. O. West: Acta Crystallogr. **B 30** (1974), 2451.
[547] G. R. Clark, D. Hall, T. N. Waters: J. chem. Soc. (London) **A 1969**, 2808.
[548] P. L. Orioli, L. Sacconi: J. Amer. chem. Soc. **88** (1966), 277.
[549] V. W. Day, M. D. Glick, J. L. Hoard: J. Amer. chem. Soc. **90** (1968), 4803.
[550] D. Hall, S. V. Sheat-Rumball, T. N. Waters: J. chem. Soc. (London) **A 1968**, 2721.
[551] J. E. Davies, B. M. Gatehouse: Acta Crystallogr. **B 29** (1973), 1934, 2651.
[552] M. N. Akhtar, A. J. Smith: Acta Crystallogr. **B 29** (1973), 275.
[553] J. E. Davies, B. M. Gatehouse: Acta Crystallogr. **B 28** (1972), 3641.
[554] M. Calligaris, G. Nardin, L. Randaccio: J. chem. Soc., Dalton Transact. **1972**, 2003.
[555] P. D. Cradwick, M. E. Cradwick, G. G. Dodson, D. Hall, T. N. Waters: Acta Crystallogr. **B 28** (1972), 45.
[556] A. P. Gardner, B. M. Gatehouse, J. C. B. White: Acta Crystallogr. **B 27** (1971), 1505.
[557] L. A. Lindblom, W. P. Schaefer, R. E. Marsh: Acta Crystallogr. **B 27** (1971), 1461.
[558] E. N. Baker, D. Hall, T. N. Waters: J. chem. Soc. (London) **A 1970**, 400, 406.
[559] Y. Kushi, T. Hosoo, H. Kuroya: J. chem. Soc., Chem. Commun. **1970**, 397.
[560] M. Gerloch, E. D. McKenzie, A. D. C. Towl: J. chem. Soc. (London) **A 1969**, 2850.
[561] D. Hall, R. H. Sumner, T. N. Waters: J. chem. Soc. (London) **A 1969**, 420.
[562] A. P. Gardner, B. M. Gatehouse, J. V. B. White: J. chem. Soc., Chem. Commun. **1968**, 694.
[563] T. P. Cheeseman, D. Hall, T. N. Waters: J. chem. Soc. (London) **A 1966**, 685.
[564] E. N. Baker, D. Hall, T. N. Waters: J. chem. Soc. (London) **A 1966**, 680.
[565] E. R. Boyko, D. Hall, M. E. Kinloch, T. N. Waters: Acta Crystallogr. **21** (1966), 614.
[566] J. A. J. Jarvis: Acta Crystallogr. **14** (1961), 961.
[567] D. Kobelt, E. F. Paulus, W. Kunstmann: Acta Crystallogr. **B 28** (1972), 1319.
[568] N. W. Alcock, R. C. Spencer, R. H. Prince, O. Kennard: J. chem. Soc. (London) **A 1968**, 2383.
[569] A. H. G. de Mesquita, C. H. MacGillavry, K. Eriks: Acta Crystallogr. **18** (1965), 437.
[570] A. Robbins, G. A. Jeffrey, J. P. Chesick, J. Donohue, F. A. Cotton, B. A. Frenz, C. A. Murillo: Acta Crystallogr. **B 31** (1975), 2395.
[571a] M. Chao, E. Schempp, R. D. Rosenstein: Acta Crystallogr. **B 31** (1975), 2922.
[571b] A. Kvick, M. Backens: Acta Crystallogr. **B 30** (1974), 474.
[572] J. S. Sherfinski, R. E. Marsh: Acta Crystallogr. **B 31** (1975), 1073.
[573] R. Destro, T. Pilati, M. Simonetta: Acta Crystallogr. **B 31** (1975), 2883.
[574] M. Chao, E. Schempp, R. D. Rosenstein: Acta Crystallogr. **B 31** (1975), 2924.
[575] G. C. Verschoor: Nature **202** (1964), 1206.
[576] P. G. Slade, M. Raupach, E. W. Radoslovich: Acta Crystallogr. **B 29** (1973), 279.
[577] W. Bolton: Acta Crystallogr. **16** (1963), 166.
[578] B. M. Craven: Acta Crystallogr. **17** (1964), 282.
[579] G. L. Gartland, B. M. Gatehouse, B. M. Craven: Acta Crystallogr. **B 31** (1975), 203.
[580] B. Berking, B. M. Craven: Acta Crystallogr. **B 27** (1971), 1107.
[581] B. M. Craven, E. A. Vizzini: Acta Crystallogr. **B 27** (1971), 1917.
[582] C. Escobar: Acta Crystallogr. **B 31** (1975), 1059.
[583] O. Dideberg, L. Dupont, D. Pyzalska: Acta Crystallogr. **B 31** (1975), 685.
[584] L. Dupont, D. Dideberg, D. Pyzalska: Acta Crystallogr. **B 30** (1974), 2447.
[585] I-Nan-Hsu, B. M. Craven: Acta Crystallogr. **B 30** (1974), 1299.
[586] I-Nan-Hsu, B. M. Craven: Acta Crystallogr. **B 30** (1974), 998.
[587] I-Nan-Hsu, B. M. Craven: Acta Crystallogr. **B 30** (1974), 994.
[588] I-Nan-Hsu, B. M. Craven: Acta Crystallogr. **B 30** (1974), 988.
[589] G. L. Gartland, B. M. Craven: Acta Crystallogr. **B 30** (1974), 980.

[590] I-Nan-Hsu, B. M. Craven: Acta Crystallogr. **B 30** (1974), 974.
[591] I-Nan-Hsu, B. M. Craven: Acta Crystallogr. **B 30** (1974), 843.
[592] P. H. Smit, J. A. Kanters: Acta Crystallogr. **B 30** (1974), 784.
[593] M. R. Caira, G. V. Fazakerley, P. W. Linder, L. R. Nassimbeni: Acta Crystallogr. **B 29** (1973), 2898.
[594] P. P. Williams: Acta Crystallogr. **B 29** (1973), 1572.
[595] H. Wunderlich: Acta Crystallogr. **B 29** (1973), 168.
[596] G. L. Gartland, B. M. Craven: Acta Crystallogr. **B 27** (1971), 1909.
[597] B. M. Gatehouse, B. M. Craven: Acta Crystallogr. **B 27** (1971), 1337.
[598] I-Nan-Hsu, D. P. Lesser, B. M. Craven: Acta Crystallogr. **B 31** (1975), 882.
[599] L. Nassimbeni, A. Rodgers: Acta Crystallogr. **B 30** (1974), 2593.
[600] L. Nassimbeni, A. Rodgers: Acta Crystallogr. **B 30** (1974), 1953.
[601] J. Berthou, R. Rerat, C. Rerat: Acta Crystallogr. **18** (1965), 768.
[602] B. Berking: Acta Crystallogr. **B 28** (1972), 1539.
[603] B. Berking: Angew. Chem. **83** (1971), 889.
[604] R. J. McClure, B. M. Craven: Acta Crystallogr. **B 29** (1973), 1860.
[605] W. Bolton: Acta Crystallogr. **17** (1964), 147.
[606] C. Singh: Acta Crystallogr. **19** (1965), 767.
[607] B. M. Craven, Y. Mascarenhas: Acta Crystallogr. **17** (1964), 407.
[608] H. Gillier: Bull. Soc. chim. France **1965**, 2385.
[609] M. Hamelin: Acta Crystallogr. **B 28** (1972), 228.
[610] K. Morikawa, K. Torii, Y. Iitaka, M. Tsuboi: Acta Crystallogr. **B 31** (1975), 1004.
[611] E. A. Green, R. D. Rosenstein, R. Shiono, D. J. Abraham, B. L. Trus, R. E. Marsh: Acta Crystallogr. **B 31** (1975), 102.
[612] U. Thewalt, C. E. Bugg: Acta Crystallogr. **B 29** (1973), 1393.
[613] A. Rahman, H. R. Wilson: Acta Crystallogr. **B 28** (1972), 2260.
[614] G. N. Reeke, R. E. Marsh: Acta Crystallogr. **20** (1966), 703.
[615] S. W. Hawkinson: Acta Crystallogr. **B 31** (1975), 2153.
[616] H. Sternglanz, G. R. Freeman, C. E. Bugg: Acta Crystallogr. **B 31** (1975), 1393.
[617] C. L. Coulter, N. R. Cozzarelli: Acta Crystallogr. **B 30** (1974), 2176.
[618] P. Singh, D. J. Hodgson: Acta Crystallogr. **B 30** (1974), 1430.
[619] J. S. Sherfinski, R. E. Marsh: Acta Crystallogr. **B 30** (1974), 873.
[620] L. Fallon: Acta Crystallogr. **B 29** (1973), 2549.
[621] R. Destro, R. E. Marsh: Acta Crystallogr. **B 28** (1972), 2971.
[622] M. R. Truter, B. L. Vickery: J. chem. Soc. (London) **A 1971**, 2077.
[623] D. W. Green, F. S. Mathews, A. Rich: J. biol. Chemistry **237** (1962), 3573.
[624] R. Gerdil: Acta Crystallogr. **14** (1961), 333.
[625] P. Singh, D. J. Hodgson: Acta Crystallogr. **B 31** (1975), 2519.
[626] K. Hoogsteen: Acta Crystallogr. **16** (1963), 28.
[627] B. L. Kindberg, E. L. Amma: Acta Crystallogr. **B 31** (1975), 1492.
[628] D. W. Young, H. R. Wilson: Acta Crystallogr. **B 31** (1975), 961.
[629] B. D. Sharma, J. F. McConnell: Acta Crystallogr. **19** (1965), 797.
[630] D. L. Barker, R. E. Marsh: Acta Crystallogr. **17** (1964), 1581.
[631] G. A. Jeffrey, Y. Kinoshita: Acta Crystallogr. **16** (1963), 20.
[632] C. E. Bugg, U. Thewalt: Acta Crystallogr. **B 31** (1975), 121.
[633] D. L. Kozlowski, P. Singh, D. J. Hodgson: Acta Crystallogr. **B 30** (1974), 2806.
[634] P. Tougard, J.-F. Chantot: Acta Crystallogr. **B 30** (1974), 214.
[635] R. Destro, T. J. Kistenmacher, R. E. Marsh: Acta Crystallogr. **B 30** (1974), 79.
[636] U. Thewalt, C. E. Bugg, R. E. Marsh: Acta Crystallogr. **B 27** (1971), 2358.
[637] J. Iball, H. R. Wilson: Proc. Roy. Soc. Ser. **A 288** (1965), 418.

[638] E. Sletten, M. Ruud: Acta Crystallogr. **B 31** (1975), 982.
[639] T. J. Kistenmacher, T. Shigematsu: Acta Crystallogr. **B 31** (1975) 211.
[640] A. Ku Chwang, M. Sundaralingam, S. Hanessian: Acta Crystallogr. **B 30** (1974), 2273.
[641] T. J. Kistenmacher, T. Shigematsu: Acta Crystallogr. **B 30** (1974), 166.
[642] P. Prusiner, M. Sundaralingam: Acta Crystallogr. **B 28** (1972), 2142.
[643] R. F. Stewart, L. H. Jensen: J. chem. Physics **40** (1964), 2071.
[644] T. J. Kistenmacher, T. Sorrell: Acta Crystallogr. **B 31** (1975), 489.
[645] H. Mizuno, F. Fujiwara, K. Tomita: Bull. chem. Soc. Japan **42** (1969), 3099.
[646] M. R. Caira, L. R. Nassimbeni: Acta Crystallogr. **B 31** (1975), 1339.
[647] E. Sletten: Acta Crystallogr. **B 30** (1974), 1961.
[648] H. C. Mez, C. J. Donohue: Z. Kristallogr. **130** (1969), 376.
[649] H. Ringertz: Acta Crystallogr. **20** (1966), 397.
[650] D. J. Sutor: Acta Crystallogr. **16** (1963), 97.
[651] J.-P. Mornon, J. Delettré, R. Bally: Acta Crystallogr. **B 31** (1975), 2119.
[652] H. Sternglanz, C. E. Bugg: Acta Crystallogr. **B 31** (1975), 890.
[653] P. Singh, D. J. Hodgson: Acta Crystallogr. **B 31** (1975), 845.
[654] R. D. Gilardi: Acta Crystallogr. **B 29** (1973), 2089.
[655] P. Prusiner, M. Sundaralingam: Acta Crystallogr. **B 28** (1972), 2148.
[656] C. Sabelli, P. F. Zanazzi: Acta Crystallogr. **B 28** (1972), 1173.
[657] Hsing-Yie Ting, W. H. Watson, X. A. Dominguez: Acta Crystallogr. **B 28** (1972), 1046.
[658] R. Norrestam, B. Stensland, E. Söderberg: Acta Crystallogr. **B 28** (1972), 659.
[659] E. Shefter, B. E. Evans, E. C. Taylor: J. Amer. chem. Soc. **93** (1971), 7281.
[660] Y. Kushi, Q. Fernando: J. Amer. chem. Soc. **92** (1970), 91.
[661] G. J. Palenik: Acta Crystallogr. **17** (1964), 687.
[662] J. E. Fleming, H. Lynton: Canad. J. Chem. **46** (1968), 471.
[663] G. J. Palenik: Acta Crystallogr. **17** (1964), 696.
[664] L. F. Power, A. M. Tait: Inorg. nucl. chem. Letters **7** (1971), 337.
[665] J.-M. Léger, J.-P. Bideau: C. R. Acad. Sci. (Paris) **275 C** (1972), 313.
[666] R. Desiderato, J. C. Terry, G. R. Freeman: Acta Crystallogr. **B 27** (1971), 2443.
[667] B. F. Studd, A. G. Swallow: J. chem. Soc. (London) **A 1968**, 1961.
[668] P. Jose, S. Ooi, Q. Fernando: J. inorg. nuclear Chem. **31** (1969), 1971.
[669] H. Paulus, S. Göttlicher: Z. Kristallogr. **130** (1969), 267.
[670] W. De W. Horrocks, D. H. Templeton, A. Zalkin: Inorg. Chem. **7** (1968), 2303.
[671] G. E. Blank, J. Pletcher, M. Sax: Acta Crystallogr. **B 30** (1974), 2525.
[672] D. Dennis, R. H. Stanford: Acta Crystallogr. **B 29** (1973), 1053.
[673] C. J. Fritchie, R. M. Johnston: Acta Crystallogr. **B 31** (1975), 454.
[674] M. C. Kuo, J. B. R. Dunn, C. J. Fritchie: Acta Crystallogr. **B 30** (1974), 1766.
[675] J. L. Wells, B. L. Trus, R. M. Johnston, R. E. Marsh, C. J. Fritchie: Acta Crystallogr. **B 30** (1974), 1127.
[676] M. Wang, C. J. Fritchie: Acta Crystallogr. **B 29** (1973), 2040.
[677] R. Karlsson: Acta Crystallogr. **B 28** (1972), 2358.
[678] R. Norrestam: Acta Crystallogr. **B 28** (1972), 1713.
[679] R. Norrestam, O. Tillberg: Acta Crystallogr. **B 28** (1972), 1704.
[680] R. Norrestam, B. Stensland: Acta Crystallogr. **B 28** (1972), 440.
[681] R. Norrestam, M. v. Glehn: Acta Crystallogr. **B 28** (1972), 434.
[682] M. Leijonmarck, P.-E. Werner: Acta chem. scand. **25** (1971), 2273.
[683] B. L. Trus, J. L. Wells, R. M. Johnstone, C. J. Fritchie, R. E. Marsh: J. chem. Soc., Chem. Commun. **1971**, 751.
[684] M. v. Glehn, P. Kierkegaard, R. Norrestam, O. Rönnquist, P.-E. Werner: Acta chem. scand. **24** (1970), 3701.

[685] P.-E. WERNER, O. RÖNNQUIST: Acta chem. scand. **24** (1970), 997.
[686] P. KIERKEGAARD, O. RÖNNQUIST, P.-E. WERNER: Tetrahedron Letters **1969**, 4667.
[687] R. NORRESTAM, M. v. GLEHN, L. O. HAGMAN, P. KIERKEGAARD: Acta chem. scand. **23** (1969), 2199.
[688] P. KIERKEGAARD, M. LEIJONMARCK, R. NORRESTAM, L. TORBJÖRNSSON, P.-E. WERNER: Acta chem. scand. **23** (1969), 2197.
[689] R. NORRESTAM, P. KIERKEGAARD, B. STENSLAND, L. TORBJÖRNSSON: J. Chem. Soc., Chem. Commun. **1969**, 1250.
[690] N. TANAKA, T. ASHIDA, Y. SASADA, M. KAKUDO: Bull. chem. Soc. Japan **42** (1969), 1546.
[691] C. J. FRITCHIE, B. L. TRUS: J. Chem. Soc., Chem. Commun. **1968**, 1486.
[692] G. NYGJERD, J. MCALISTER, M. SUNDARALINGAM, S. MATSUURA: Acta Crystallogr. **B 31** (1975), 413.
[693] N. R. STEMPLE, W. H. WATSON: Acta Crystallogr. **B 28** (1972), 2485.
[694] I. L. KARLE, J. KARLE: Acta Crystallogr. **B 27** (1971), 1891.
[695a] S. DÄHNE, D. LEUPOLD: Angew. Chem. **78** (1966), 1029; Angew. Chem. int. Edit. **5** (1966), 984.
[695b] S. DÄHNE, S. KULPE: J. prakt. Chem., submitted.
[696] A. I. KIPRIANOV: Uspechi chimii **40** (1971), 1283.
[697] R. RADEGLIA, S. DÄHNE: Z. Chem. **13** (1973), 474.
[698] S. DÄHNE, D. LEUPOLD, R. RADEGLIA: J. prakt. Chem. **314** (1972), 525.
[699] S. DÄHNE, J. RANFT, H. PAUL: Tetrahedron Letters **1964**, 3355.
[700] S. DÄHNE, D. LEUPOLD, H. PAUL, J. RANFT: Angew. Chem. **77** (1965), 1017.
[701] S. DÄHNE, H. PAUL, D. LEUPOLD: Z. Naturforsch. **20b** (1965), 76.
[702] S. DÄHNE, H. PAUL: Chem. Ber. **97** (1964), 1625.
[703] S. DÄHNE, M. SEEBACHER: Z. Chem. **12** (1972), 141.
[704] D. LEUPOLD, S. DÄHNE: J. molecular Spectroscopy **17** (1965), 325.
[705] J. FABIAN, G. TRÖGER-NAAKE: J. prakt. Chem. **318** (1976), 801.
[706] J. FABIAN, G. TRÖGER-NAAKE: Int. J. Quantum Chem. **11** (1977), 259.
[707] S. KULPE, S. DÄHNE: Acta Crystallogr., in press.
[708] S. KULPE: J. prakt. Chem. **316** (1974), 353.
[709] S. KULPE, S. DÄHNE: Tetrahedron Letters **1968**, 2591.
[710] H. P. TROMMSDORFF, D. BOURDEAUX, D. MENTZAFOS: C. R. Acad. Sci. Paris **271 C** (1970), 45.
[711] D. RABINOVICH, G. M. J. SCHMIDT: J. chem. Soc. (London) **1964**, 2030.
[712] J. TROTTER: Acta Crystallogr. **13** (1960), 86.
[713a] H. IRNGARTINGER, R.-D. ACKER, W. REBAFKA, H. A. STAAB: Angew. Chem. **86** (1974), 705.
[713b] J. TROTTER: J. chem. Soc., Perkin Transact. II, **1973**, 476.
[714] E. K. ANDERSEN: Acta Crystallogr. **22** (1967), 196.
[715] K. J. VAN WEPEREN, G. J. VISSER: Acta Crystallogr. **B 28** (1972), 338.
[716] B. REES: Acta Crystallogr. **B 26** (1970), 1311.
[717] B. REES: Acta Crystallogr. **B 26** (1970), 1304.
[718] B. REES: Acta Crystallogr. **B 26** (1970), 1298.
[719] B. REES: Acta Crystallogr. **B 26** (1970), 1292.
[720] S. S. C. CHU, G. A. JEFFREY, T. SAKURAI: Acta Crystallogr. **15** (1962), 661.
[721] E. K. ANDERSEN: Acta Crystallogr. **22** (1967), 201.
[722] G. B. JENSEN, E. K. ANDERSEN: Acta Crystallogr. **17** (1964), 243.
[723] R. G. DELAPLANE, J. A. IBERS: Acta Crystallogr. **B 25** (1969), 2423.
[724] E. K. ANDERSEN, I. G. K. ANDERSEN: Acta Crystallogr. **B 31** (1975), 384.
[725] E. K. ANDERSEN: Acta Crystallogr. **22** (1967), 188.

[726] E. K. Andersen: Acta Crystallogr. 22 (1967), 191.
[727 a] K. J. Palmer, R. Y. Wong, L. Jurd: Acta Crystallogr. B 29 (1973), 1509.
[727 b] H. P. Klug: Acta Crystallogr. 19 (1965), 983.
[728] E. K. Andersen: Acta Crystallogr. 22 (1967), 204.
[729] E. K. Andersen, I. G. K. Andersen: Acta Crystallogr. B 31 (1975), 379.
[730] S. Kulpe, S. Dähne: Angew. Chem. 78 (1966), 639.
[731] S. J. Rettig, J. Trotter: Canad. J. Chem. 53 (1975), 777.
[732] S. Kulpe: Z. Chem. 11 (1971), 466.
[733] T. Ito, T. Sakurai: Acta Crystallogr. B 29 (1973), 1594.
[734] K. Wallenfels, W. Draper: Tetrahedron (London) 20 (1964), 1889.
[735] D. R. Carter, J. W. Turley, F. P. van Remoortere, F. P. Boer: Acta Crystallogr. B 29 (1973), 1971.
[736] J. E. Weidenborner, E. Fahr, M. J. Richter, K.-H. Koch: Angew. Chem. 85 (1973), 229.
[737] J. W. L. van Oijen, C. Romers: Acta Crystallogr. 20 (1966), 169.
[738] E. Subramanian, J. Trotter: J. chem. Soc. (London) A 1970, 622.
[739] J. Silverman, J. Stam-Thole, C. H. Stam: Acta Crystallogr. B 27 (1971), 1846.
[740] A. E. Švec, Ja. Ja. Blejbelis, R. P. Šibaeva, L. O. Atovmjan: Žurnal strukturnoj chimii 13 (1972), 745.
[741] K. Hafner, K. F. Bangert, V. Orfanos: Angew. Chem. 79 (1967), 414.
[742] H. L. Ammon, M. Sundaralingam, J. M. Stewart: Acta Crystallogr. B 25 (1969), 336.
[743] J. Gaultier, C. Hauw: Acta Crystallogr. 18 (1965), 179.
[744] C. K. Prout, E. E. Castellano: J. chem. Soc. (London) A 1970, 2775.
[745] J. Gaultier, C. Hauw, J. Housty, M. Schvoerer: C. R. Acad. Sci. (Paris) 273 C (1971), 956.
[746] J. Gaultier, C. Hauw: Acta Crystallogr. 19 (1965), 919.
[747] J. Gaultier, C. Hauw: Acta Crystallogr. 18 (1965), 604.
[748] J. Gaultier, C. Hauw: Acta Crystallogr. 19 (1965), 580.
[749] C. Courseille, J. Gaultier, C. Hauw, M. Schvoerer: C. R. Acad. Sci. (Paris) 270 C (1970), 687.
[750] C. Courseille, S. Jeoffre, M. Schvoerer: C. R. Acad. Sci. (Paris) 273 C (1971), 1633.
[751] J. Gaultier, C. Hauw: Acta Crystallogr. B 25 (1969), 419.
[752] J. Gaultier, C. Hauw: Acta Crystallogr. 21 (1966), 694.
[753] J. Gaultier, C. Hauw: Acta Crystallogr. 20 (1966), 620.
[754] J. Gaultier, C. Hauw: Acta Crystallogr. 19 (1965), 585.
[755] D. Chasseau, G. Bravic: C. R. Acad. Sci. (Paris) 272 C (1971), 377.
[756] A. Aime, J. Gaultier, C. Hauw: Acta Crystallogr. B 26 (1970), 1597.
[757] D. Chasseau, J. Gaultier, C. Hauw: C. R. Acad. Sci. (Paris) 270 C (1970), 1452.
[758] P. D. Cradwick, D. Hall: Acta Crystallogr. B 27 (1971), 1990.
[759] C. Pascard-Billy: Bull. Soc. chim. France 1962, 2282, 2293, 2299.
[760] C. Pascard-Billy: Acta Crystallogr. 15 (1962), 519.
[761] P. Srivastava: Indian J. Physics Proc. Indian Assoc. Cultivat. Sci. 35, [44] (1961), 640.
[762] K. Lonsdale, H. J. Milledge, K. El Sayed: Acta Crystallogr. 20 (1966), 1.
[763] A. Prakash: Acta Crystallogr. 22 (1967), 439.
[764] M. Ehrenberg: Acta Crystallogr. 22 (1967), 482.
[765] L. A. Četkina, G. A. Golder, G. S. Ždanov: Kristallografija 8 (1963), 194.
[766] L. A. Četkina, G. A. Golder: Kristallografija 12 (1967), 404.
[767] L. A. Četkina: Kristallografija 12 (1967), 42.
[768] J. Guilhem: Acta Crystallogr. 23 (1967), 330.
[769] D. Hall, C. L. Nobbs: Acta Crystallogr. 21 (1966), 927.
[770] M. Bailey, C. J. Brown: Acta Crystallogr. 22 (1967), 493.
[771] G. H. Stout, L. H. Jensen: Acta Crystallogr. 15 (1962), 451.

[772] A. Prakash: Z. Kristallogr. **122** (1965), 272.

[773] N. L. Klimasenko, L. A. Četkina, S. L. Ginzburg, M. G. Neuhaus, E. M. Smeljanskaja: Žurnal strukturnoj chimii **14** (1973), 108.

[774] G. H. Stout, V. F. Stout, M. J. Welsh: Tetrahedron (London) **19** (1963), 667.

[775] G. H. Stout, T. Shun Lin, I. Singh: Tetrahedron (London) **25** (1969), 1975.

[776] M. Röhrl, W. Hoppe: Chem. Ber. **103** (1970), 3502.

[777] M. Bailey, C. J. Brown: Acta Crystallogr. **22** (1967), 392.

[778] M. Bailey, C. J. Brown: Acta Crystallogr. **22** (1967), 488.

[779] K. Ogawa, H.-J. Scheel: Z. Kristallogr. **130** (1969), 405.

[780] M. Bailey: Acta Crystallogr. **8** (1955), 182.

[781] H. L. Ammon: Acta Crystallogr. **B 26** (1970), 1690.

[782] H. Koyama, H.-J. Scheel, F. Laves: Naturwissenschaften **53** (1966), 700.

[783] W. Lüttke, M. Klessinger: Chem. Ber. **97** (1964), 2342.

[784] W. Lüttke, D. Hunsdiecker: Chem. Ber. **99** (1966), 2146.

[785] P. W. Sadler: J. org. Chemistry **21** (1956), 316; J. Amer. chem. Soc. **78** (1956), 1251.

[786] L. Fitjer, W. Lüttke: Chem. Ber. **105** (1972), 937.

[787] E. Wille, W. Lüttke: Angew. Chem. **83** (1971), 853.

[788] H. Hermann, W. Lüttke: Chem. Ber. **101** (1968), 1715.

[789] W. Lüttke, H. Hermann, M. Klessinger: Angew. Chem. **78** (1966), 638.

[790] M. Klessinger, W. Lüttke: Tetrahedron (London) **19**, Suppl. 2 (1963), 315.

[791] M. Klessinger: Tetrahedron (London) **22** (1966), 3355.

[792] E. A. Gribova, G. S. Ždanov, G. A. Golder: Kristallografija **1** (1956), 53.

[793] H. v. Eller: Bull. Soc. chim. France **1955**, 1433.

[794] H. v. Eller: Bull. Soc. chim. France **1955**, 1438.

[795] H. v. Eller: Bull. Soc. chim. France **1955**, 1444.

[796] H. Pandraud: Acta Crystallogr. **15** (1962), 1131.

[797] G. H. Goldschmidt, F. J. Llewellyn: Acta Crystallogr. **3** (1950), 294.

[798] H. J. A. Hermann, H. L. Ammon, R. E. Gibson: Tetrahedron Letters **1969**, 2559.

[799] H. v. Eller-Pandraud: Acta Crystallogr. **13** (1960), 936.

[800] H. Pandraud: Acta Crystallogr. **14** (1961), 901.

[801] R. H. Thomson: Angew. Chem. **86** (1974), 355.

[802] F. Giordano, L. Mazzarella, G. Prota, C. Santacroce, D. Sica: J. chem. Soc. (London) **C 1971**, 2610.

[803] W. M. Macintyre, M. S. Werkema: J. chem. Physics **40** (1964), 3563.

[804] C.-H. Wong, R. E. Marsh, V. Schomaker: Acta Crystallogr. **17** (1964), 131.

[805] I. L. Karle, K. Britts, S. Brenner: Acta Crystallogr. **17** (1964), 1506.

[806] H. J. Lindner, B. v. Gross: Chem. Ber. **107** (1974) 598.

[807] H. I. Irngartinger, H. Rodewald: Angew. Chem. **86** (1974), 783.

[808] A. Skancke: Acta chem. scand. **22** (1968), 3239.

[809] N. C. Baenziger, J. J. Hegenbarth, D. G. Williams: J. Amer. chem. Soc. **85** (1963), 1539.

[810] N. C. Baenziger, D. G. Williams: J. Amer. chem. Soc. **88** (1966), 689.

[811] M. D. Glick, G. L. Downs, L. F. Dahl: Inorg. Chem. **3** (1964), 1712.

[812] M. D. Glick, L. F. Dahl: Inorg. Chem. **5** (1966), 289.

[813] W. Haberditzl: Angew. Chem. **78** (1966), 277.

[814] W. Haberditzl, R. Havemann, P. Grzegorzewski: Z. physik. Chem. (Leipzig) **217** (1961), 91.

[815] J. H. Fuhrhop: Angew. Chem. **86** (1974), 363.

[816] L. E. Webb, E. B. Fleischer: J. chem. Physics **43** (1965), 3100.

[817] M. J. Hamor, T. A. Hamor, J. L. Hoard: J. Amer. chem. Soc. **86** (1964), 1938.

[818] E. B. FLEISCHER, C. K. MILLER, L. E. WEBB: J. Amer. chem. Soc. **86** (1964), 2342.
[819] D. M. COLLINS, W. R. SCHEIDT, J. L. HOARD: J. Amer. chem. Soc. **94** (1972), 6689.
[820] J. L. HOARD: Science (Washington) **174** (1971), 1295.
[821] W. R. SCHEIDT, J. L. HOARD: J. Amer. chem. Soc. **95** (1973), 8281.
[822] W. R. SCHEIDT, J. A. CUNNINGHAM, J. L. HOARD: J. Amer. chem. Soc. **95** (1973), 8289.
[823] T. SAKURAI, K. YAMAMOTO, N. SEINO, M. KATSUTA: Acta Crystallogr. **B 31** (1975), 2514.
[824] L. J. RADONOVICH, A. BLOOM, J. L. HOARD: J. Amer. chem. Soc. **94** (1972), 2073.
[825] D. M. COLLINS, R. COUNTRYMAN, J. L. HOARD: J. Amer. chem. Soc. **94** (1972), 2066.
[826] A. B. HOFFMAN, D. M. COLLINS, V. W. DAY, E. B. FLEISCHER, T. S. SRIVASTAVA, J. L. HOARD: J. Amer. chem. Soc. **94** (1972), 3620.
[827] D. M. COLLINS, J. L. HOARD: J. Amer. chem. Soc. **92** (1970), 3761.
[828] E. B. FLEISCHER: J. Amer. chem. Soc. **85** (1963), 146.
[829] T. A. HAMOR, W. S. CAUGHEY, J. L. HOARD: J. Amer. chem. Soc. **87** (1965), 2305.
[830] E. F. MEYER: Acta Crystallogr. **B 28** (1972), 2162.
[831] J. L. HOARD, M. J. HAMOR, T. A. HAMOR, W. S. CAUGHEY: J. Amer. chem. Soc. **87** (1965), 2312.
[832] R. C. PETTERSEN: Acta Crystallogr. **B 25** (1969), 2527.
[833] W. S. SHELDRICK, J.-H. FUHRHOP: Angew. Chem. **87** (1975), 456.
[834] S. W. HAWKINSON, E. B. FLEISCHER: Inorg. Chem. **8** (1969), 2402.
[835] E. B. FLEISCHER, E. KLEM: Inorg. Chem. **4** (1965), 637.
[836] W. HOPPE, G. WILL, J. GASSMANN, H. WEICHSELGARTNER: Z. Kristallogr. **128** (1969), 18.
[837] J. D. DUNITZ, E. F. MEYER: Helv. chim. Acta **54** (1971), 77.
[838] M. CURRIE, J. D. DUNITZ: Helv. chim. Acta **54** (1971), 98.
[839] M. DOBLER, J. D. DUNITZ: Helv. chim. Acta **54** (1971), 90.
[840] M. K. FRIEDEL, B. F. HOSKINS, R. L. MARTIN, S. A. MASON: J. chem. Soc., Chem. Commun. **1970**, 400.
[841] D. ROGERS, R. S. OSBORN: J. chem. Soc., Chem. Commun. **1971**, 840.
[842] K. UKEI: Acta Crystallogr. **B 29** (1973) 2290.
[843] C. J. BROWN: J. chem. Soc. (London) **A 1968**, 2488.
[844] L. H. VOGL, A. ZATKIN, D. H. TEMPLETON: Science (Washington) **151** (1966), 569.
[845] A. GIEREN, W. HOPPE: J. chem. Soc., Chem. Commun. **1971**, 413.
[846] T. L. CHOCJANOVA, S. I. KUZNECOV, T. A. BABUŠKINA: Žurnal structurnoj chimii **9** (1968), 713.
[847] J. L. DE BOER, A. VOS: Acta Crystallogr. **B 28** (1972), 839.
[848] J. L. DE BOER, A. VOS: Acta Crystallogr. **B 28** (1972), 835.
[849] J. L. DE BOER, A. VOS, K. HUML: Acta Crystallogr. **B 24** (1968), 542.
[850] J. TANAKA, N. SAKABE: Acta Crystallogr. **B 24** (1968), 1345.
[851] R. CHANDRASEKARAN: Acta Crystallogr. **B 25** (1969), 369.
[852] S. A. CHAWDHURY, A. HARGREAVES, S. H. RIZVI: Acta Crystallogr. **B 24** (1968), 1633.
[853] S. A. CHAWDHURY, A. HARGREAVES, R. A. L. SULLIVAN: Acta Crystallogr. **B 24** (1968), 1222.
[854] D. HLAVATÁ: Acta Crystallogr. **B 27** (1971), 1483.
[855] K. TOMAN, D. OČENÁŠKOVA, K. HUML: Acta Crystallogr. **22** (1967), 32.
[856] L. L. MERITT, E. D. SCHROEDER: Acta Crystallogr. **9** (1956), 801.
[857] P. MURRAY-RUST: Acta Crystallogr. **B 31** (1975), 1771.
[858] G. J. ASHWELL, S. C. WALLWORK, S. R. BAKER, P. I. C. BERTHIER: Acta Crystallogr. **B 31** (1975), 1174.
[859] T. SUNDARESAN, S. C. WALLWORK: Acta Crystallogr. **B 28** (1972), 3065.
[860] T. SUNDARESAN, S. C. WALLWORK: Acta Crystallogr. **B 28** (1972), 2474.
[861] J. H. RUSSELL, S. C. WALLWORK: Acta Crystallogr. **B 28** (1972), 1527.

[862] J. H. Russell, S. C. Wallwork: Acta Crystallogr. B 27 (1971), 2473.
[863] C. K. Prout, P. Murray-Rust: J. chem. Soc. (London) A 1969, 1520.
[864] J. H. Russell, S. C. Wallwork: Acta Crystallogr. B 25 (1969), 1691.
[865] S. S. Basson, L. D. C. Bok, J. G. Leipoldt: Acta Crystallogr. B 25 (1969), 579.
[866] M. B. Ferrari, A. B. Corradi, G. G. Fava, C. G. Palmieri, M. Nardelli, C. Pelizzi: Acta Crystallogr. B 29 (1973), 1808.
[867] R. H. Fenn: J. chem. Soc. (London) A 1969, 1764.
[868] I. M. Procter, F. S. Stephens: J. chem. Soc. (London) A 1969, 1248.
[869] R. E. Long, R. A. Sparks, K. N. Trueblood: Acta Crystallogr. 18 (1965), 932.
[870] M. Konno, Y. Saito: Acta Crystallogr. B 31 (1975), 2007.
[871] M. Konno, Y. Saito: Acta Crystallogr. B 30 (1974), 1294.
[872] G. R. Anderson, C. J. Fritchie: 2nd National Meeting, Society for Applied Spectroscopy, San Diego USA, 14—18 Okt., 1963.
[873] A. Hoekstra, T. Spoelder, A. Vos: Acta Crystallogr. B 28 (1972), 14.
[874] T. Sundaresan, S. C. Wallwork: Acta Crystallogr. B 28 (1972), 3507.
[875] C. J. Fritchie, P. Arthur: Acta Crystallogr. 21 (1966), 139.
[876] H. Kobayashi, Y. Ohashi, F. Marumo, Y. Saito: Acta Crystallogr. B 26 (1970), 459.
[877] H. Kobayashi, T. Danno, Y. Saito: Acta Crystallogr. B 29 (1973), 2693.
[878] T. J. Kistenmacher, T. E. Phillips, D. O. Cowan: Acta Crystallogr. B 30 (1974), 763.
[879] A. W. Hanson: Acta Crystallogr. 19 (1965), 610.
[880] A. W. Hanson: Acta Crystallogr. B 24 (1968), 768.
[881] R. H. Colton, E. D. Henn: J. chem. Soc. (London) B 1970, 1532.
[882] R. M. Williams, S. C. Wallwork: Acta Crystallogr. B 24 (1968), 168.
[883] F. Iwasaki: Acta Crystallogr. B 27 (1971), 1360.
[884] H. Kobayashi: Bull. chem. Soc. Japan 47 (1974), 1346.
[885] M. Konno, Y. Saito: Acta Crystallogr. B 29 (1973), 2815.
[886] T. Sundaresan, S. C. Wallwork: Acta Crystallogr. B 28 (1972), 1163.
[887] H. Kobayashi, F. Marumo, Y. Saito: Acta Crystallogr. B 27 (1971), 373.
[888] T. H. Herbstein: Perspectives in Structural Chemistry Vol. IV p. 166, J. Wiley and Sons, Inc., Publ., New York 1971.
[889] T. Sundaresan, S. C. Wallwork: Acta Crystallogr. B 28 (1972), 491.
[890] G. Briegleb: Elektronen-Donator-Akzeptor-Komplexe, Springer Verlag Berlin, Göttingen, Heidelberg, 1961.
[891a] J. C. A. Boeyens, F. H. Herbstein: J. physic. Chem. 69 (1965), 2153.
[891b] C. K. Prout, D. J. Wright: Angew. Chem. 80 (1968), 688.
[892] C. J. Fritchie: Acta Crystallogr. 20 (1966), 892.
[893] D. Chasseau, J. Gaultier, C. Hauw: C. R. Acad. Sci. (Paris) 274 C (1972), 1434.
[894] A. N. Bloch, Do. Cowan, K. Beckgaard: Physic. Rev. Letters 34 (1975), 1561.
[895] Y. Ohashi: Acta Crystallogr. B 29 (1973), 2863.
[896] H. Tsuchiya, F. Marumo, Y. Saito: Acta Crystallogr. B 29 (1973), 659.
[897] J. L. de Boer, A. Vos: Acta Crystallogr. B 24 (1968), 720.
[898] K. Yakushi, I. Ikemoto, H. Kuroda: Acta Crystallogr. B 27 (1971), 1710.
[899] K. Yakushi, I. Ikemoto, H. Kuroda: Acta Crystallogr. B 29 (1973), 2640.
[900] N. Tachikawa, K. Yakushi, H. Kuroda: Acta Crystallogr. B 30 (1974), 2770.
[901] K. Yakushi, I. Ikemoto, H. Kuroda: Acta Crystallogr. B 30 (1974), 1738.
[902] K. Yakushi, I. Ikemoto, H. Kuroda: Acta Crystallogr. B 30 (1974), 835.
[903] H. Kobayashi: Acta Crystallogr. B 30 (1974), 1010.
[904] I. Goldberg, U. Shmueli: Acta Crystallogr. B 29 (1973), 421, 432, 440.
[905] I. Ikemoto, K. Chikaishi, K. Yakushi, H. Kuroda: Acta Crystallogr. B 28 (1972), 3502.
[906] R. Richter, J. Kaiser, J. Sieler, L. Kutschabsky: Acta Crystallogr. B 31 (1975), 1642.

[907] A. Kutoglu: Acta Crystallogr. **B 29** (1973), 2891.
[908] J. H. Enemark, J. A. Ibers: Inorg. Chem. **7** (1968), 2636.
[909] I. Leichert, J. Weiss: Acta Crystallogr. **B 31** (1975), 2709.
[910] I. Leichert, J. Weiss: Acta Crystallogr. **B 31** (1975), 2877.
[911] D. Ginderow: Acta Crystallogr. **B 31** (1975), 1092.
[912] L. P. Battaglia, A. B. Corradi, C. G. Palmieri, M. Nardelli, M. E. Vidoni Tani: Acta Crystallogr. **B 30** (1974), 1114.
[913] R. H. Bowers, C. V. Banks, R. A. Jacobson: Acta Crystallogr. **B 28** (1972), 2318.
[914] H. Endres, H. J. Keller, W. Moroni, J. Weiss: Acta Crystallogr. **B 31** (1975), 2357.
[915] C. D. Stout, M. Sundaralingam, G. Hung-Yin Lin: Acta Crystallogr. **B 28** (1972), 2136.
[916] J. T. Veal, W. E. Hatfield, D. J. Hodgson: Acta Crystallogr. **B 29** (1973), 12.
[917] H. S. Preston, C. H. L. Kennard: J. chem. Soc. (London) **A 1969**, 2682.
[918] H. S. Preston, C. H. L. Kennard: J. chem. Soc. (London) **A 1969**, 1956.
[919] D. E. Williams: Acta Crystallogr. **B 29** (1973), 96.
[920] R. Pariser: J. chem. Physics **25** (1957), 1112.
[921] G. S. Pawley: Acta Crystallogr. **18** (1965), 560.
[922] J. M. Robertson, H. M. M. Shearer, G. A. Sim, D. G. Watson: Acta Crystallogr. **15** (1962), 1.
[923] O. Bastiansen, J. L. Derissen: Acta chem. scand. **20** (1966), 1319.
[924] H. Ammon, M. Sundaralingam: J. Amer. chem. Soc. **88** (1966), 4794.
[925] L. Fallon, H. L. Ammon, A. G. Anderson, J. O. Currie, R. A. Labar: Acta Crystallogr. **B 30** (1974), 531.
[926] A. W. Hanson: Acta Crystallogr. **19** (1965), 19.
[927] M. R. Churchill, P. H. Bird: Inorg. Chem. **7** (1968), 1545.
[928] M. R. Churchill, P. H. Bird: Inorg. Chem. **7** (1968), 1793.
[929] M. R. Churchill, P. H. Bird: J. Amer. chem. Soc. **90** (1968), 3241.
[930] P. H. Bird, M. R. Churchill: Inorg. Chem. **7** (1968), 349.
[931] M. R. Churchill, J. Wormald: Inorg. Chem. **8** (1969), 716.
[932] M. R. Churchill: Inorg. Chem. **6** (1967), 190.
[933] M. R. Churchill, K. Gold, P. H. Bird: Inorg. Chem. 8 (1969), 1956.
[934] H. J. Lindner: Chem. Ber. **102** (1969), 2464.
[935] H. J. Lindner: J. chem. Soc. (London) **B 1970**, 907.
[936] H. J. Lindner: Chem. Ber. **102** (1969), 2456.
[937] R. Allmann, T. Debaerdemaeker: Crystal Struct. Commun., in press.
[938] R. Allmann, T. Debaerdemaeker, A.-R. Ferwanah, W. Pressler, C. Reichardt: Chem. Ber., **109** (1976), 3005.
[939] W. D. Erhardt, H. L. Ammon: Tetrahedron Letters **1975**, 3997.

Bei Fragen zur Produktsicherheit wenden Sie sich bitte an:
If you have any questions regarding product safety,
please contact:

Walter de Gruyter GmbH
Genthiner Straße 13
10785 Berlin
productsafety@degruyterbrill.com